PLANT DISEASES AND THEIR BIOLOGICAL CONTROL

PLANT DISEASES AND THEIR BIOLOGICAL CONTROL

Iqbal Hasan

RAJAT PUBLICATIONS
New Delhi

Plant Diseases and their Biological Control
by Iqbal Hasan

This edition is published by Rajat Publications
4740/23, Ansari Road Daryaganj,
New Delhi - 110002

ISBN: 978-81-7880-424-8

Edition 2026

Printed and bound in India

RAJAT
PUBLICATIONS

4740/23, Ansari Road, Daryaganj,
New Delhi- 110002 (India)
Tel:- +91-11-23267924
Mob:- +91-9811484973, +91-9811105389
E-mail: rajat_publication@yahoo.com

Contents

Preface

The day Humans stared performing agricultural practices during prehistoric times; control of plant diseases has remained a challenge. In spite of large-scale application of agrochemicals and pesticides during recent decades, the number of plant diseases is still beyond control. Every year agricultural damages due to pests and pathogens cause enormous economic losses and exports are restricted owing to poor quality. Adverse short- and long term effects of agrochemicals on the ecosystem have prompted exploration of natural alternatives for the bio-control of plant diseases on a sustainable basis.

Safety and environmental concerns have forced the withdrawal of a number of synthetic agrochemicals from the market. As biological control alone is the logical path forward for a sustainable ecosystem, there is no viable alternative except to give it a fair chance to prove its ability. The use of biological control strategies offers several advantages over chemical control, since it is economical, self-perpetuating, and usually free from residual side effects. Biological approaches for bio-control envisages principles, concepts, mechanisms, and current practices using predator's Antagonistic Microbes, Rhizosphere Micro flora, and Genetic Engineering.

Prof. Iqbal Hasan

Chapter 1

Plant Disease Concepts

Plant pathology or phytopathology is the science dealing with plant diseases and their control. Plant pathologists study plant diseases caused by fungi, bacteria, viruses, nematodes, and parasitic plants. They also study plant disorders caused by nutrient imbalances, air pollution, and other unfavorable growing conditions.

Plant diseases have had profound effects on mankind through the centuries as evidenced by Biblical references to the blasting and mildew of plants. The Greek philosopher Theophrastus (370-286 B.C.) was the first to describe maladies of trees, cereals, and legumes that we currently classify as leaf scorch, rots, scab, and cereal rust. The Romans were also aware of rust diseases of their grain crops. They celebrated the holiday of Robigalia when sacrifices of reddishcolored dogs and cattle were made in an attempt to appease the rust god Robigo. With the invention of the microscope in the 17th century, fungi and bacteria associated with plants were investigated. In 1665, Robert Hooke published the first illustration of rust on a rose leaf. Advances in the study of diseases were hampered by the widely held belief in the theory of spontaneous generation. This theory, held by most people in the mid-18th century, considered pathogenic or disease causing microorganisms as products of disease rather than causes of disease.

Epidemics of late blight of potato devastated Ireland in 1845 and 1846. These epidemics dramatized the effect of plant diseases on mankind. Tragically, these epidemics caused famine and death for over a million people. Between 1845 and 1860, death and migration accounted for the loss of nearly one-third of Ireland's population. In 1861, a German botanist, Anton De Bary, proved that a fungus (*Phytophthora infestans*) was the causal agent of late blight of potato.

This was a milestone in the study of plant diseases since it showed that a fungus was indeed the cause of a plant disease rather than an organism simply associated with the disease. Two years later, Louis Pasteur proposed his germ theory of disease that finally disproved the theory of spontaneous generation and changed the way modern science investigated the diseases of all living organisms.

DISEASE CONCEPTS

WHAT IS A DISEASE

There are many ways to define what a plant disease is. However, simply put, plant diseases involve profound changes within the host that cause a disruption of normal plant function. A good working definition of a healthy plant is one that can carry out its physiological functions to the best of its genetic ability. Diseases are among the most important factors that can significantly diminish growth and yield, or reduce the usefulness of a plant or plant product. Healthy or normal plants develop and function to the maximum of their genetic potential. However, when plants are adversely affected by continuous irritation by a disease-causing agent, which interferes with normal development and functioning, plants are considered to be diseased. This broad definition excludes injury or damage such as mechancial injury (e.g., lawn mower or weed-eater injury to trees); deer, rodent, and bird damage; hail damage; and lightning injury.

In addition to reduction in growth, yield, and economic or aesthetic value of a plant or plant product, diseases may lead to the death of the whole plant or destruction of the entire crop under conditions favorable for the disease. Diseases may interfere with absorption and translocation of water and nutrients from the soil to the various parts of the plant, may reduce the photosynthetic efficiency of the plant parts, may interrupt the translocation of photosynthetic products through the plant, or may interfere with the reproduction and storage of food reserves in the plant.

Diseases in plants are caused by either living (biotic, parasitic, or infectious) agents called pathogens, or nonliving (abiotic, nonparasitic, or noninfectious) environmental factors. Plant diseases may also be grouped by the causal agent involved (fungal diseases, bacterial diseases, viral diseases, nematode diseases, etc.), the plant part affected (root diseases, seedling diseases, leaf diseases, stem diseases, flower diseases, fruit diseases, tuber diseases, etc.), or the types of symptoms (damping-off, wilts, leaf spots, cankers, blights, galls, root knots, mosaics, storage rots, etc.).

Symptoms of Diseases

Symptoms are the visible reactions of a plant to a disease and may suggest a causal agent. A sampling of disease symptoms might include wilting, necrosis, abnormal coloration, defoliation, fruit drop, abnormal cellular growth, or stunting of the infected plant. However, it is important to remember that different disease agents can cause similar symptoms on the same host. An equally important point to remember is that insect feeding can also cause disease-like symptoms on plants.

Signs of Diseases

Signs are the visible parts of the pathogen or its products seen on the host that can be used to identify the pathogen. Examples of common disease signs include: the white coating of mycelium visible on powdery mildew-infected leaves, mushroom growth on a tree limb, droplets of bacterial ooze running down a fruit tree twig, nematode cysts on plant roots, or dark fungal fruiting bodies visible in leaf lesions.

Causal Agents of Disease

A pathogen is any organism that can cause a disease. Pathogens cause infectious diseases that can spread from an infected plant to a healthy plant. Pathogens that cause infectious diseases include bacteria, fungi, viruses, nematodes, and parasitic plants. Plant disease can also be caused by noninfectious or nonliving factors. Causes of disease by nonliving factors include unfavorable growing conditions, mineral deficiencies, and air pollution.

Pathogens that cannot be cultured apart from their host are classified as obligate parasites. Pathogens that can be cultured apart from their hosts on artificial media are called nonobligate parasites. In general, obligate parasites only attack very specific host plants, whereas nonobligate parasites typically have a wider range of plants they can infect. Some pathogens are restricted to a single plant species, while others infect a single plant genus. Still others attack a large number of hosts from many plant genera.

There are also several levels of parasitism that pathogens can have with their hosts. When a pathogen is capable of infecting a plant, the plant is considered susceptible to that pathogen. If a pathogen cannot infect a plant, then the plant is considered immune to that pathogen. Plants can vary in their response to pathogens from high resistance (very little disease development), to partial resistance (moderate disease development), or high susceptibility (severe disease development). Pathogens can vary in their degree of virulence on a susceptible plant ranging from highly virulent (causing severe disease symptoms) to weakly virulent (causing less disease).

Inoculum and Pathogen

Dissemination Inoculum is any part of the pathogen that can cause infection. Examples of inoculum include fungal spores, bacterial cells, virus particles, or nematode eggs. Inoculum that survives the winter and causes the original or primary infection in the spring is called primary inoculum. Secondary inoculum causes additional infections throughout the growing season.

Inoculum is sometimes present at the site where a plant is grown and can also be introduced from an outside source.

Inoculum already present at a plant site includes soil pathogens or pathogens that overwinter on perennial weeds. Introduced inoculum includes infected plant material such as infected seeds, wind-blown fungal spores, and inoculum transmitted by insects.

Inoculum can be disseminated passively by wind, rain, and man. Inoculum can also be disseminated actively by insects and nematodes or fungal zoospores swimming through water in the soil toward plant roots. Only a fraction of any pathogen's inoculum will ever land on a susceptible host. The vast majority of inoculum lands on material that cannot be infected. Most pathogens produce a tremendous surplus of inoculum.

Pathogen Survival

Pathogens in temperate climates must have a way of overwintering when their host plants are dormant or absent. In perennial plants, pathogens can survive in infected plant parts such as roots, bulbs, stems, and bud scales. Annual plants, however, die at the end of the growing season and pathogens must survive in insects, seeds, or as resistant spores.

Factors Affecting Disease Occurrence

Diseases in plants are an exception rather than a rule. Three factors, called the disease triangle must coincide for a plant to become diseased: the host, the pathogen, and the environment. The interaction between these three factors with time determines the occurrence and severity of a disease. For the disease to occur, the following conditions must be met:

- The host plant must be of a susceptible species or cultivar at the right stage of development (susceptible host).
- The pathogen must be of a virulent race or strain and must be present in sufficient numbers (inoculum potential). The presence of appropriate vectors or other agents of dispersal is also necessary.
- The environmental (atmospheric and soil) conditions such as temperature, humidity, rainfall,

wind, moisture, light, soil type, texture and pH, density of planting, aeration, and nutritional status (mineral deficiency or excess) must be favorable for disease development.

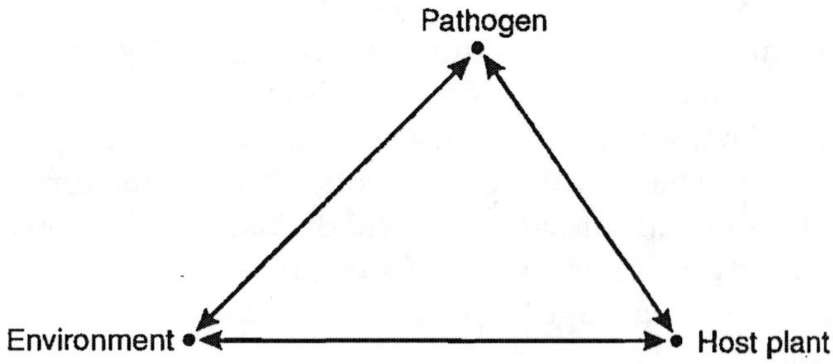

Fig. Plant Disease Triangle.

- Understanding the various aspects of the host, the pathogen, the environment, as well as their interaction is essential to implement an effective disease management strategy.

DISEASES CAUSED BY LIVING AGENTS

FUNGI

Commonly known as molds, fungi (singular=fungus) are mostly microscopic organisms that have bodies (mycelium) composed of multi-cellular, thread-like, branched filaments (hyphae) and reproductive structures called spores. Since they do not possess chlorophyll, fungi depend on either dead organic matter or living plants for their growth and reproduction. Some fungi produce vitamins and antibiotics that are useful to us.

A few fungi, like some types of mushrooms and morels, are edible. On the other hand, some fungi thrive on living plants, drawing their nutrition from them and sometimes producing toxins that cause disease and death of the plants they infect. These are called plant pathogenic fungi. A majority of diseases in plants are caused by fungi. Some examples commonly encountered in home gardens and landscape trees are: brown rot of cherries, apple scab, black spot of rose, snapdragon rust, corn smut, powdery mildew of rose, peach leaf curl, sycamore anthracnose, early blight of potato, Verticillium wilt of tomato, damping- off, and root rot of vegetables.

Bacteria and Phytoplasmas

Bacteria (singular=bacterium) and phytoplasmas (formerly known as mycoplasmas or mycoplasma-like organisms) are microscopic, single-celled organisms that cause some of the most destructive diseases in plants. Some bacteria, like those that induce nodulation in leguminous plants, are beneficial to plants because they fix nitrogen from the air into the root nodules in a form that the host plant can utilize for its growth.

Phytoplasmas are a type of bacteria that lack distinct cell walls. Under favorable conditions, bacteria reproduce very rapidly and can cause serious damage in a short period of time. Bacterial pathogens are spread by wind-splashed rain, insects, contaminated seed, or implements. Bacterial diseases are relatively difficult to control because there are very few chemicals that are effective against them.

Some commonly encountered bacterial diseases are: crown gall of rose, grape, apple, cherry, and other ornamental plants; fire blight of apple and pear; soft rot of potato; ring rot of potato; and aster yellows phytoplasma on carrots, tomatoes, onions, lettuce, etc.

Viruses

Viruses are infectious agents so small they must be observed through an electron microscope. Particles of these viruses may be in the form of rods, spheres, or threads. They are composed mainly of a nucleic acid core surrounded by a protein coat. Viruses can multiply only in a living host cell and can often spread systemically throughout the infected plant.

Viruses can be transmitted from infected to healthy plants mechanically, through grafts, and by contaminated propagating material. Viruses can also be transmitted by certain organisms, referred to as vectors. In addition to insects (primarily aphids, white flies, leafhoppers, and beetles), virus vectors include mites, nematodes, and fungi in the soil. Viral diseases are not controlled by pesticide chemicals. Examples of viral diseases are: curly top of tomato, bean, cucurbits, etc.; potato leaf roll; bean common mosaic; and rose mosaic.

Nematodes

Nematodes are microscopic roundworms that live in soil as well as water, and survive as eggs or cysts. Most of them are saprophytes, but some infect living plants and cause diseases. Most plant parasitic nematodes feed on the underground parts of the plants (roots, tubers, bulbs, etc.) causing lesions or root knots. However, a few nematodes also affect the buds, leaves, flowers, and stems of plants. Nematodes spread through contaminated planting material (tubers, seedlings, etc.), manure, soil, water, machinery, and implements. Some nematodes are vectors of plant viruses.

Some examples of plant parasitic nematodes are: root knot nematodes of tomato, potato, beans, and many other plants; root lesion nematodes of corn and potatoes, cyst nematode of sugarbeets; stubby root nematode of corn; stem and bulb nematodes of onion; and foliar nematode of chrysanthemum.

Parasitic Higher (Flowering) Plants

Several flower- and seed-producing plants live as parasites on other plants (host plants), deriving their nutrition from them and adversely affecting the host plant's growth and yield. Dodder (also know as strangleweed and devil's hair), for example, parasitizes several garden plants such as potatoes and carrots. It produces orange or yellow vine strands that entwine the stems and other plant parts from which it draws its nutrition through tube-like structures it introduces into the host tissue. Dodder produces abundant seeds that ensure its propagation and spread. Another example of a parasitic plant is dwarf mistletoes on pines.

DISEASES CAUSED BY NONLIVING AGENTS

A variety of environmental and cultural factors can cause

diseases in plants. Since these diseases occur in the absence of pathogens, they do not spread from a diseased plant to a healthy plant.

HIGH OR LOW TEMPERATURES

When plants or plant parts are exposed to high temperatures for prolonged periods, symptoms of scorching or scalding may develop. Some examples are: sunburn or scorching of leaves and sunscald of fruits (e.g., apples, tomatoes, peppers, and melons). Similarly, low temperatures, like frost or freeze, can damage the exposed or sensitive organs (buds, flowers, young fruits, etc.) or may kill the entire plant.

Examples include: southwest-side damage to trunks of apple trees; frost damage to blossoms and young apple fruits; russet ring (caused by frost) on apple and pear fruits; winter injury to trees; and frost damage to tomatoes, beans, potatoes, etc.

HIGH OR LOW SOIL MOISTURE

Too much moisture due to excessive watering, poor drainage, ponding, or flooding may cause plants to turn yellow and be stunted. Potted indoor plants, for example, may show poor development or root rots. Seedlings are vulnerable to damping-off caused by soilborne pathogens under these conditions. In some indoor or greenhouse plants (e.g., geraniums, begonias) growing under warm, humid atmospheric conditions and excessive soil moisture, a condition known as edema (small, wart-like rusty, corky bumps) can develop on the underside of the leaves, and on the stems. At the other extreme, low moisture or drought conditions can lead to poor development, wilting, and death of plants.

HIGH OR LOW LIGHT INTENSITY

High light intensity is usually not a problem but low light conditions, especially for indoor plants, can lead to etiolation (spindly or lanky plant growth with chlorotic yellow foliage).

Lack of Aeration or Low Oxygen Supply

Low aeration can deprive plant roots of adequate oxygen

and can adversely affect their development or even kill the plant. Inadequate oxygen supply during the storage of potato tubers can lead to the development of a condition called blackheart, the browning and death of internal tuber tissue.

Air Pollution

Certain chemcals, such as ozone, sulfur dioxide, and nitrogen dioxide are released into the air from factories, power plants, and automobile exhausts. These chemicals can accumulate in the atmosphere in sufficient concentration to cause damage to plants. Ozone damage appears in the form of mottling, chlorosis, spots, and bleaching of young leaves.

This is common in certain regions of the country where there is a high ozone concentration in smog. For example, ozone damage is frequently found on the leaves of beans, petunias, and grapes. Some of the air pollutants responsible for acid rain cause damage to vegetation in certain regions. In Idaho, however, plant damage due to air pollution is not common.

Nutrient Deficiencies

Plants require several major (nitrogen, phosphorus, potassium, calcium, magnesium, and sulfur) and minor (iron, boron, copper, zinc, etc.) elements for normal growth. Deficiency or lack of any of these essential nutrients results in disease symptoms in the plant. Specific symptoms depend on the plant species and the deficient nutrient. If not corrected, a prolonged acute deficiency of essential nutrients can lead to death of the plant. Common examples of nutrient deficiencies are: nitrogen deficiency in beans, iron deficiency in peaches, zinc deficiency in apple trees, and calcium deficiency in apple fruit (bitter pit). In the home garden, the common blossom-end rot of tomato fruit is caused, in part, by calcium deficiency.

Mineral Toxicity

Presence of excessive available amounts of certain minerals in the soil can lead to mineral toxicity to the plants. The extent of injury depends on the mineral, its concentration, and the species of the plant. Excessive amounts of sodium salts in the soil can lead to high pH and to alkali injury (e.g., alkali injury to apple). Plants growing in acidic soils can be injured by aluminum or manganese toxicity.

Unfavorable Soil pH

Although many plants can grow in a rather wide range of soil pH, plants growing in soils with unfavorable pH usually show poor growth and mineral deficiency or toxicity symptoms.For example, iron deficiency symptoms are very common in plants growing in high pH soils in Idaho. Under conditions of high soil pH, iron in the soil becomes unavailable to the plant, thus inducing interveinal chlorosis and yellowing of leaves. The plant may die if the condition remains uncorrected for a prolonged time.

Pesticide Toxicity

Some pesticides, if improperly used, can cause serious damage to plants. For example, if wettable sulfur is sprayed (for powdery-mildew control) on a very hot day (above 90°F), it will result in injury to the plant phytotoxicity). However, the most common type of chemical injury to plants is due to soil residues or spray drift of herbicides. Examples of pesticide toxicity are: 2,4-D damage to beans and tomatoes, dicamba (Banvel) damage to vegetables and trees, and glyphosate (Roundup) damage to fruit trees. Some herbicides used as soil sterilants may leave the soil unsuitable for any plant growth for several years.

Improper Cultural Practices

Any cultural practice done in the wrong way or at the wrong time can result in significant damage to plants. Injury can result from improper amounts of chemical fertilizer or pesticide or improper chemical mixes in the spray tank. Root pruning can result from excessively deep cultivation; distorted and twisted roots can result from pot-bound conditions of a plant. African violet leaves sprinkled with very cold water develop rings and ring-like patterns that resemble symptoms caused by some viruses.

DIAGNOSING PLANT DISEASE PROBLEMS

IMPORTANCE OF DIAGNOSIS

Diagnosing plant problems can involve considerable detective work. Sometimes there is insufficient information and other times, the primary cause of a problem is hidden by more obvious, but less-important, problems. Success in diagnosing plant problems depends on how much we know about the host plant, about the plant problems in general, and the quality of information obtained from the client.

For example, 10 tomato plants all similarly damaged, are brought. All have yellow leaves, stunted growth, and very few feeder roots. We learn from questioning the grower that he applied one-half of a 20- pound bag of 10-10-10 fertilizer to a tomato plot that measures 60 square feet. The grower put 10 pounds of fertilizer on 60 square feet, which translates to a rate of 166 pounds per 1,000 square feet. This is almost 10 times the normal rate of 20 pounds per 1,000 square feet of a 10-10-10 fertilizer. The grower's fertilization rate is enough to kill fine feeder roots. The diagnosis is damage to the roots caused by overfertilization.

BASIC STEPS IN REACHING A DIAGNOSIS

- Identify the plant—The better our plant identification skills the faster we will be able to diagnose a problem. Most references on plant pests and diseases are organized by plant, so knowing the plant is the essential first step in using many reference books.
- What is normal? Our familiarity with the normal appearance and cultural requirements of the plant will enable us to differentiate normal changes from symptoms of a problem.
- What is the problem? To make a disease diagnosis, we need to know: the pattern of distribution of the diseased plants or plant parts, the plant species or cultivar involved, the site where the plant is growing (field, orchard, garden, greenhouse, inside the house, etc.), and previous crop history of the site.

For example, uniform damage to many species in an area, to all plants on one side of the field or garden, or to all shoots on one side of the tree indicates that the cause may be an abiotic factor. Also, if the damage is well demarcated in a garden or in a plant, it may suggest that some abiotic factor is involved. On the other hand, if there is evidence of progressive spread of the disease from an initial focus to other plants of the same cultivar or species or to different parts of the plant, it may indicate that an infectious agent is involved.

- Examine the plant and note symptoms and signs —For a presumptive diagnosis of diseases in plants, look for the symptoms and signs of the disease. The characteristic internal or external alterations of a plant in response to a disease-causing agent are called symptoms (leaf spot, necrosis, blight, canker, wilt, lesion, gall, witches' broom, rot, chlorosis, mosaic, etc.). Sometimes, the pathogen that causes the disease produces its own characteristic growth or structures on the diseased plant that are of diagnostic value. These are referred to as signs of the disease (mold, mildew, sclerotia, mushrooms, conks, etc.).
- Tentative diagnosis—Based on our knowledge of the plant and information, formulate a tentative diagnosis. This will help us in examination of the plant and assist in collecting relevant information.
- Double-check the diagnosis—Once we have arrived at a diagnosis, unless it is an obvious diagnosis, double-check it. Ask other master gardeners or extension educators for their opinions.
- Types of plant disease diagnosis—Verbal descriptions by a telephone call or evaluation of a sample provide the most common diagnostic opportunities.

There will be common, familiar problems, such as powdery mildew of apples, when a little information easily leads to a correct diagnosis. In other cases, it will be very difficult to make a diagnosis over the telephone and it may be necessary to evaluate a sample. Much of our diagnositic work as a master gardener will be done with plant samples.

PRINCIPLES OF PLANT DISEASE CONTROL

Plant disease control aims at preventing or reducing the amount of damage or economic loss. Most plant disease-control measures are aimed at preventing or protecting plants from the disease rather than curing the plant after it is diseased. By the time the disease symptoms appear, it is often too late to reverse the damage caused by the disease agent. In some cases, acceptable control measures are not available to halt the disease, short of plant removal. Correct diagnosis of the disease is important in order to implement a control strategy to prevent or reduce the incidence of the disease in the next crop cycle.The various controls for diseases can be classified as exclusion strategies, inoculum reduction methods, use of disease-resistant cultivars, chemical control, and biological controls. Integrated disease- management strategies may utilize any or all of these methods.

EXCLUSION OF THE PATHOGEN

Some destructive plant diseases do not occur in our country, state, or region. To prevent import and introduction of these pathogens national, regional, or state regulations, known as quarantines, are put into force. Some examples are the customs and plant health inspection services at airports and check points on highways. In addition, national and state laws regulate the conditions under which certain crops may be grown and distributed between states and countries. We cannot import several types of plant material from abroad without specific permission and inspection by the appropriate quarantine authority.

An example of a state quarantine is in relation to white rot of onion and garlic. White rot is a very destructive disease of onion and garlic. Once introduced in the soil, the pathogen can survive for over 20 years and there are no satisfactory chemical control measures. At present, our commercial onion growing areas in southern Idaho are free from this disease.

Therefore, the State Department of Agriculture in Idaho restricts the importation of onion and garlic bulbs and other related plants for planting in southern Idaho.

Restriction of bean seed from outside of Idaho prevents the introduction of two serious bacterial diseases: halo blight and common blight. Homeowners can keep some serious plant pathogens out of their yards and gardens by planting only certified, disease-free planting material (seeds, tubers, bulbs, seedlings, nursery stock, etc.). For example, a homeowner can avoid several virus diseases and ring rot of potato by planting certified disease- free tubers.

ERADICATION OR REDUCTION OF THE PATHOGEN POPULATION

Serious disease damage to a crop or garden patch can be prevented or reduced by pulling out and destroying the first plant or the first few plants that show the disease symptoms (roguing). This prevents the spread of the pathogen to other healthy plants (e.g., elimination of the first bean plants with mosaic symptoms).

Since some of the root-infecting pathogens remain viable in the soil even after the crop is harvested, avoid planting the same or similar susceptible crop in that part of the garden for the next 2 to 3 years. An appropriate crop rotation in the garden is very essential to reduce the damage caused by root rot and wilt-causing pathogens and nematodes (e.g., Fusarium root rot and wilt, Verticillium wilt, and nematodes on tomato). Eliminating the infected leaves and diseased or dead branches, as well as using clean garden tools and similar methods of sanitation, will prevent the spread and build up of disease-causing organisms. Other cultural practices, such as growing plants on raised beds and using composted tree bark in the planting medium for containerized nursery stock, help reduce the damage caused by certain soilborne pathogens (e.g., damping-off, root rots, and wilts). Proper irrigation management can prevent some diseases (e.g., collar rot of apple). Using the sun's heat during the summer months to reduce some of the soilborne pathogens can be successful in certain regions. In this process, called soil solarization, the soil is cultivated to a fine texture and deeply watered.The moist soil is covered with a clear polyethylene sheet. The sheet's edges are buried to create an airtight environment.

The moist soil below the polyethylene mulch heats up and is slowly "cooked." Under sunny weather conditions for at least 4 weeks, this process reduces or eliminates several soilborne pathogens. Marigolds interplanted with nematodesusceptible crops produce substances in the soil that are toxic to plant parasitic nematodes and thereby reduce the nematode damage to the crop plant. Controlling insects with insecticides can reduce the spread of some of the diseases caused by viruses that are transmitted by insects (e.g., potato leaf roll virus spread by aphids).

USE OF DISEASE-RESISTANT CULTIVARS

Where available, use of disease-resistant cultivars is the most cost-effective, safe, easy, and environmentally desirable option for the gardener. In fact, for some diseases (such as curly top of tomato), resistant cultivars offer the only practical control option available. Varieties resistant to one or more diseases are available in several vegetables and fruits (e.g., tomato cultivars with resistance to Fusarium wilt, Verticillium wilt, nematodes, and tomato mosaic virus; and apple cultivars resistant to fire blight).

CHEMICAL PROTECTION OF PLANT

The most common method of direct protection of plants against plant pathogens is through the use of chemicals. Chemicals used for control of fungal diseases are called fungicides, those that control bacterial diseases are called bactericides, and those that control nematodes are called nematicides.No chemicals at this time are effective against diseases caused by viruses. Certain chemicals, used to control soilborne pathogens, are called fumigants. These are highly volatile chemicals that are toxic to all the living organisms (biocides), including insects, weed seeds, fungi, bacteria, and nematodes. An example of a soil fumigant is methyl bromide.

Fumigants need extreme care in handling and application, and therefore, can only be used by certified pesticide applicators.

Based on the mode of action, pesticide chemicals can be classified as follows:

Protectants: These chemicals, when applied to the plants, remain on the plant surface and prevent spore germination and infection of the plant by the pathogen (e.g., sulfur, captan).

Systemics: These chemicals are absorbed by plants and are translocated to other plant parts (e.g., Benlate, Ridomil).

Eradicants: These chemicals can eradicate the pathogen from plant tissue after the infection has occurred; that is, after the penetration by and establishment of the pathogen inside the host tissue (e.g., Rally).

Based on the type of chemical (active ingredient), these pesticides can also be classified as copper compounds, sulfur compounds, dithiocarbamates, benzimidazoles, antibiotics, etc. Disease-control chemicals can be applied as seed treatments, foliar sprays, dusts, fumigants, wound paints, dips, or through irrigation water (chemigation). The effectiveness of chemical control measures is dependent upon our understanding of the disease cycle, host susceptibility, timing, mode of application, coverage, and choice of the appropriate product.

Timing of the application and thorough coverage of the plant surface are very important for effective disease control. For best results, most chemicals are applied before the pathogen infects the host. Also, most chemical sprays have to be applied several times at regular intervals for best disease control.

Always follow the label directions and precautions. Development of pathogen strains resistant to chemicals or antibiotics (e.g., streptomycin-resistant strains of apple fire blight pathogen) can lead to these products becoming ineffective in disease control.

BIOLOGICAL CONTROL

The strategy for biological control of plant diseases involves the use of antagonistic microorganisms before or after infection takes place. There has been successful control of crown gall with strain K84 of *Agrobacterium radiobacter*, which produces an antibiotic specific against crown gall bacterium. Commercial biological control agents are available as seed treatments and soil amendments to protect plants against soilborne pathogens.

Currently, the bacteria *Bacillus subtilis* and *Pseudomonas* spp. and the fungi *Gliocladium virens* and *Trichoderma* spp. are the organisms with the most applications in biological control strategies. There is tremendous research interest in developing new biological tools for plant disease control, but research-based options are limited at this time. Genetic engineering of biocontrol microorganisms has sparked wider environmental concerns, which will limit the speed of this new technology's use in plant disease control.

INTEGRATED DISEASE MANAGEMENT

An effective, economical, and sustainable disease management strategy should incorporate all the available approaches of a disease management programme. It should include the relevant preventive and control measures appropriate for the crop, such as selection of planting site, selection of the most adapted and disease-resistant variety, crop rotation, pathogen-free planting material, seed treatment, appropriate planting date, planting depth and density, irrigation and fertilizer use, weed control, sanitation, timely pesticide applications, proper harvesting, and handling and storage of the produce.

The integrated disease management measures selected should be effective (should control the disease), economical (should result in an economic return), and sustainable (should be environmentally sound).

Chapter 2

Classification of Plants Diseases

Some knowledge of the habits of the organism causing a disease is usually necessary in order successfully to combat it and prevent its ravages. Those diseases caused by powdery mildew fungi (which are surface infestations) can be cured. Practically all others must be prevented. Fungi attacking parts of plants above ground are usually disseminated by means of spores. Water is often necessary to liberate the spores from the fungus proper, and is nearly always necessary to permit spore germination and infection of other plants. Heavy dew sometimes furnishes sufficient moisture, but prolonged drizzling rains are more favorable. For this reason a fungicide, in order to be effective against such parasites, must be applied before the rain. If it is going to rain to-morrow, spray to-day. But how know whether it is going to rain?

This can best be told from a study of the United States weather maps, which are printed and distributed from the many weather stations, or else appear in the daily papers. Storm periods, indicated by a "low" barometer, travel quite regularly from west to east, and are usually accompanied or followed by rain. This can be determined by noting the amount of precipitation, if any, in the wake of the storm. Local conditions are often a factor to be considered. A few minutes' study of the weather map each day will soon make one reasonably efficient in predicting the weather.

It is unfortunate that a definite system of naming plant diseases has not been formulated. Diseases of plants of a similar nature should bear the same common name. The term "blight" is commonly used for many kinds or forms of diseases. It might well be restricted to bacterial diseases like fire-blight of pear or bean blight.

When some definite system of naming diseases is adopted, it is likely that a tabulation of methods of control will be somewhat simplified, for if the term " blight" is restricted to bacterial diseases of the nature of pear blight, it will be understood that certain control measures, such as spraying, will not be effective. At present, each case must be considered separately, and the popular names are used. These names are followed by the technical botanical name of the organism causing the disease, in italics, and this by a brief description of the disease, the most prominent symptom being mentioned first.

GENERAL UNCLASSIFIED DISEASES

Damping-Off

A term applied to the decay of young seedlings or cuttings at or near the surface of the ground. The trouble is due to the action of various organisms, especially Pythium deBaryanum, Phytophthora cactorum, Rhizoctonia sp., etc. Wet soil, confined atmosphere, and crowded plants are conducive to damping-off. ***Control:*** Steam-sterilize seed or cutting beds. Sterilize nursery seed beds with formalin, using 1 gallon of 1 per cent solution to the square foot, i.e. 1 pint of formalin in 12-15 gallons of water.

Cedema or Dropsy

A disorder of various plants under glass, as tomatoes, violets, geraniums, which have insufficient sunlight, stimulating temperature and soil, and too much moisture. It has also been observed on twigs of the apple. It is usually indicated by elevated corky or spongy points or masses, much resembling fungous injury. The leaves curl. The only remedy is to improve conditions under which the plants are grown.

Smuts of Cereals

Practically every cereal is attacked by a specific smut

fungus, and most of them by two perfectly distinct species. These smuts are confined to a single species of cereal, and never cross from one to another. Some of the smuts produce a loose black spore-mass (loose smuts), while in others (covered smuts) the seed coat of the grain is not affected, so that the smut is not detected until the grain is broken open.

The most important difference to be noted, however, is the method of wintering. In some the spores adhere to the surface of the seed and infect the young seedling plant at the time of germination, while in the other case the spores fall upon the blossoms and grow down into the seed directly, there lying dormant until the seed is planted.

Control: The treatment is very different in the two cases. If the spore is on the surface of the seed, it may be killed with formalin; but if the seed is infected internally, a different treatment is necessary. The formalin treatment is very simple and inexpensive. Select a clean place on the barn floor, and heap the seed grain upon it.

Make a solution of formalin at the rate of 1 pint of formalin to 50 gallons of water. Use as many gallons of this solution as there are bushels of grain to treat. Shovel the grain over, and at the same time spray the formalin over with a sprinkling pot. Shovel over twice, and then cover two hours or over night with blankets or canvas. Spread out the grain to dry.

Make allowance for swelling of the seed at the rate of one peck per acre. When the infection is internal, the hot water process of treatment must be resorted to. Obtain a reliable thermometer, and make arrangements to keep a quantity of water at perfectly uniform temperature. Soak the seed in water at ordinary temperature for five to seven hours. Then place it in small loose sacks or wire baskets containing not more than a half peck each, and allow to drain.

Provide two tubs or vats, of 30 or 40 gallons capacity, which can be heated, or provide in addition an iron kettle for heating a quantity of water. Heat the water in the two vats to the temperature indicated. Immerse the drained sacks of seed in tub 1 to remove the chill, then suspend in tub 2 for the indicated length of time. Keep the temperature of tub 2 constant by applying heat or adding small amounts of boiling water. Treat for the indicated time, remove, and dry.

CLASSIFIED DISEASES

Barley

Covered Smut (Ustilago hordei): The covering is thin and easily broken, and when old may resemble loose smut.

Seedling Infection

Control: Make a solution of formalin at the rate of 1 pint of formalin to 50 gallons of water. Use as many gallons of this solution as there are bushels of grain to treat. Shovel the grain over, and at the same time spray the formalin over with a sprinkling pot. Shovel over twice, and then cover two hours or over night with blankets or canvas. Spread out the grain to dry.

Loose Smut (Ustilago nuda): The smutted heads are loose and black from the first.

Flower Infection

Control: Heat the water in the two vats to the temperature indicated. Immerse the drained sacks of seed in tub 1 to remove the chill, then suspend in tub 2 for the indicated length of time. Keep the temperature of tub 2 constant by applying heat or adding small amounts of boiling water. Treat for the indicated time, remove, and dry.

The temperature of tub 2 should be 127° F., and the seed should be left in fifteen minutes. If the temperature of tub 2 varies slightly from 127°, the length of treatment should be lengthened or shortened accordingly as the temperature is below or above that desired.

In no case should the temperature go above 129° or below 124° F. This treatment will also be effective for covered smut.

Oats

Loose Smuts (Ustilago avenae and Ustilago levis): Both characteristic loose smuts, and both seedling infection.

Control: Make a solution of formalin at the rate of 1 pint of formalin to 50 gallons of water. Use as many gallons of this solution as there are bushels of grain to treat. Shovel the grain over, and at the same time spray the formalin over with a sprinkling pot. Shovel over twice, and then cover two hours or over night with blankets or canvas. Spread out the grain to dry.

Wheat

Stinking Smut or Bunt (Tilletia foetens): Can be detected in the field by the flaring of the beards, in the bin by the peculiar fetid odor and by breaking open the kernels. The seed coat remains intact.

Seedling Infection

Loose Smut (Ustilago tritici): Characteristic loose smut of the head appearing at blossoming time.

Control: Make a solution of formalin at the rate of 1 pint of formalin to 50 gallons of water. Use as many gallons of this solution as there are bushels of grain to treat. Shovel the grain over, and at the same time spray the formalin over with a sprinkling pot. Shovel over twice, and then cover two hours or over night with blankets or canvas. Spread out the grain to dry.

Flower Infection

Control: Heat the water in the two vats to the temperature indicated. Immerse the drained sacks of seed in tub 1 to remove the chill, then suspend in tub 2 for the indicated length of time. Keep the temperature of tub 2 constant by applying heat or adding small amounts of boiling water. Treat for the indicated time, remove, and dry.

The temperature of tub 2 should be 129° F., and the seed should be left in ten minutes. If the temperature of tub 2 should go above 129° or fall below 126° the length of treatment should be diminished or increased accordingly. In no case should the temperature go above 131° or below 124°F.

Storage rots (Penicillium expansum and P. italicum): These two organisms are responsible for much of the rot appearing in storage or transportation. The former is the common one on apples, the latter on oranges and lemons. These organisms are not able to enter through an unbroken surface, but are dependent upon cracks, bruises, scab spots, etc. *Control:* Avoid puncturing the skins with shears or fingernails, handle and pack with care to prevent bruises, and spray to prevent scab spots. Store at a temperature of 32°. In making long distance shipments, pre-cool the car and ship under ice.

DISEASES OF DIFFERENT PLANTS OR CROPS

Alfalfa

Leaf Spot (Pseudopeziza medicaginis): Small black spots on the leaves. Causes the leaves to turn yellow and fall.

Control: Frequent close mowing usually holds the disease in check.

Dodder (Cuscuta epithymum): A tangled mat of yellow threads entwining the alfalfa stems. Usually appears in spots in the field and spreads from these points. Is easily spread by the rake, and especially in seed. Dodder is not a fungus, but a specialized parasitic plant of the morning-glory family.

Control: As soon as discovered, cover the infested spot with straw and oil and burn. Screen the alfalfa seed to remove seed of dodder. Make a screen 12 inches square by 3 inches deep with a 20 X 20 mesh wire-cloth made of No. 34 steel wire. Sift each half pound of seed vigorously for one half minute.

Apple

Blight: The same disease as Pear Blight.

Bitter-rot or Ripe-rot (Glomerella rufomaculans): Produces a browning and drying of the fruit. Progressing in concentric rings from a central point. Attacks nearly mature fruit. Also occurs on limbs, where it produces a canker scarcely distinguishable.

Control: Trim out all cankers early in the spring, and remove all mummied apples from the trees. In addition to the sprayings for apple scab, make three, four, or five sprayings with bordeaux mixture, 3-3-50, according to the severity of the disease and the character of the summer as regards rainfall.

Black-rot of fruit: Fruit stage of the New York apple-tree canker disease, which see.

Blotch (Phyllosticta solitaria): Attacks fruit, twigs, and leaves. Blotches a quarter of an inch or more in diameter appear on the fruit. These often coalesce, and the fruit often cracks deeply. Scurfy cankers are formed on the twigs while very small; circular spots a quarter of an inch in diameter are formed on the leaves. Ben Davis is especially susceptible.

Control: Careful pruning to remove cankered twigs. Spray as for apple scab and bitter rot.

Control: Smooth cankers in bark of trunk and limbs usually indicate blight, rough ones New York apple-tree canker.

Collar-rot: A dead area in the bark near the ground; often girdles the tree. Cause not known. May be started in some cases by the fire-blight organism, in others by winter injury. Common on King, Baldwin, and Ben Davis.

Remedy: As soon as noticed, cut away dead bark and wood to the living healthy tissue. Swab the wound with a solution of corrosive sublimate, 1:1000, and paint over with a lead paint which is free from turpentine. Slit the callus on the edge from year to year to make it spread faster, and keep dead wood well protected with paint.

Apple-tree Canker (Sphceropsis malorum): The fungus causing the disease attacks limbs, causing roughened cankers and often girdling the limb; attacks leaves, causing a reddish brown leaf-spot, and on the fruit produces a black rot. Abundant on Twenty Ounce.

Control: Remove and burn old cankers. Clean out and disinfect small cankers as for collar-rot. Soak old limbs well with spray mixture when spraying for scab. Spraying as for apple scab usually controls black rot of fruit, though in the Ozark region a late spraying may be advisable for leaf-spot. Cultivate thoroughly.

Powdery Mildew (Sphcerotheca leucotricha): Attacks

nursery stock, covering the leaves with a grayish white, powdery mildew. Also on leaves and twigs of new growth in the orchard, often causing the leaves to fall.

Remedy: Lime-sulfur, 1-40, as applied for scab is a specific.

Rust (Gymnosporangium macropus): A bright yellow rust appearing on the young leaves and fruit. Enfeebles the whole tree and produces one-sided fruits. It is known that one stage in the cycle of the fungus is the cedar apple, which occurs on the red cedar. Apples are always infected from the cedar, never from apple to apple.

Control: Destroy red cedars in the neighborhood, also wild apples and hawthorns. Spray thoroughly in the spring as for scab.

Scab (Venturia incequalis): Olive green, brownish or blackish scab-like spots on leaves and fruit. Arrests growth, and often causes distortion. In severe cases may make the leaves and young fruit fall. Makes leaves susceptible to spray injury. The fungus is known to be dependent upon weather conditions. The fungus winters regularly on the dead fallen leaves. In the milder climate of Virginia, the fungus may winter on the twigs.

Control: Rake and burn leaves, or plow under very early (before blossom buds open). Spray with lime-sulfur 32° Beaume, 1-40, or bordeaux, 3-3-50:

- When blossom buds show pink, but before they open;
- When the majority of petals have fallen;
- Tthree weeks after b depending upon the weather;
- If a late attack is feared, spray thoroughly before the fall rains begin.

Asparagus

Rust (Puccinia asparagi): A rust of the tops, which is often so severe as to kill them, thus interfering with root development.

Control: Three weeks after cutting stops dust the young tops with dry sulfur at the rate of 11/2 sacks of sulfur per acre. This should be done very early in the morning while the dew is still on, and only on a dewy morning. In a month or less make another application, using 2 sacks of sulfur per acre. The sulfur must go on in a dusty, smoky cloud and form a covering over all the growth. Flowers of sulfur is more satisfactory for this work, and is less expensive in the long run. Dusting machines may be obtained on the market.

Bean

Anthracnose or Pod-spot (Colletotrichum lindemuthianum): Reddish-brown scab-like spots appearing on stems, pods, and veins of leaves, particularly on yellow-podded snap beans. The fungus grows through the pod and into the young bean seed. It lies dormant in the seed, and becomes active when the bean is planted.

Control: Select pods which are free from the spots and save the seed for planting. Such seed will grow a clean crop. If disease appears in the garden, it can be controlled by thoroughly hand spraying the vines from beneath as well as above, repeating the operation every ten days as long as necessary.

Blight (Bacterium phaseoli): A bacterial disease. Causes large, papery spots on leaves and watery spots on pods.

Control: Select pods which are free from the spots and save the seed for planting. Such seed will grow a clean crop. If disease appears in the garden, it can be controlled by thoroughly hand spraying the vines from beneath as well as above, repeating the operation every ten days as long as necessary.

Bean, Lima

Blight (Phytophthora phaseoli): Attacks the pods in August and September, covering them with a white, felted coating. It also attacks shoots and leaves.

Control: Spray with bordeaux, 4–4–50, beginning about August first, and making applications at intervals of ten days or two weeks.

Beet. Heart-rot (Phoma betae): Leaves appear spotted late in July, then wilt, and finally a dry heart rot appears.

Control: Destroy infected plants. Practice long rotation. Treat seed with formalin, 1 pint in 30 gallons of water.

Leaf-spot (Cercospora beticola): Ashen gray spots with reddish borders occurring on leaves. In advanced stages, leaf becomes much cracked and torn.

Control: Spray with bordeaux mixture, 4–4–50, at frequent intervals.

Scab (Oospora scabies): Fungus produces a scabby patch on the root. The same disease as potato scab.

Control: Avoid planting beets after potatoes for several years.

Crown-gall or Root-gall (Bacterium tumefaciens): A bacterial disease which soon ruins the bushes.

Treatment: Plow up and burn all bushes in a diseased patch. Plant clean roots in a new place.

Cabbage

Club-root or Club-foot (Plasmodiophora brassicae): A contorted swelling of the roots of cabbage in the seed bed or field, preventing the plant from heading and causing it to assume a sickly colour. Occurs on many allied plants—turnips, cauliflower, Brussels sprouts, chard, radish, wild mustard, etc.

Control: Destroy affected seedlings. Rotate crops, and do not follow with other susceptible crops. Keep down weeds on which disease occurs. Lime the soil at least eighteen months before planting to cabbage, using at the rate of two tons of quicklime to the acre.

Black-rot (Bacillus campestre): The bacteria causing this disease get into the sap tubes, turn them black, and cause the leaves to drop, thus preventing heading.

Control: Practice crop rotation. Soak the seed for fifteen minutes in a solution of mercuric chloride, one tablet in a pint of water.

Carnation

Rust (Uromyces caryophyllinus): Produces brown, powdery pustules on stems and leaves.

Control: Take cuttings only from healthy plants. Pick off

diseased leaves. Spray once in two weeks with a solution of copper sulfate, 1 pound to 20 gallons. Keep water from leaves, and grow the plants at as low temperature as is compatible with best development.

Stem-rot (Rhizoctonia and Fusarium): The former produces a sudden wilting of the plant, and the stems are soon dead and dry. The latter produces a slow rot of the heart, one branch dying at a time. The treatment is the same.

Control: In the field change the location every year. In the greenhouse sterilize the soil with steam.

Celery

Early Leaf-blight (Cercospora apii): A spotting and eventual blighting of the leaves early in the summer. Begins in the seed-bed. It is favored by hot weather, either wet or dry.

Control: Spray with ammoniacal copper carbonate, 5-3-50, beginning in the seed bed and keeping the new growth covered throughout the season.

Late Blight (Septoria petroselini var. apii): A fungous disease, appearing late in the season, causing a blight of the foliage, and often destructive after the celery is stored.

Control: As above, except that spraying should be continued up to harvesting time. In either case, the disease is practically controlled by growing the plants under half shade.

Cherry

Brown-rot (Sclerotinia fructigena): Attacks flowers, leaves, and fruit. The flowers die and decay, the leaves become discolored with irregular brown spots, and the fruit rots on the tree. Attacks also peaches, plums, and apples.

Control: Spray with bordeaux mixture, 4-4-50, or lime-sulfur, 1-40,

- Just before the blossom buds open;
- Just after the blossoms fall;
- Make one or two more applications at intervals of ten days.

Powdery Mildew (Podosphaera oxycanthae): Attacks

leaves and twigs, often causing defoliation. Serious on nursery stock. Spraying as for brown rot usually controls this trouble. If it appears, spray with lime sulfur, 1-40, or dust heavily with powdered sulfur.

Leaf-spot (Cylindrosporium padi): A fungous disease in which the leaves become thickly spotted with reddish or brown spots and fall prematurely. The spots often drop out, leaving shot holes.

Control: Spray with lime sulfur, 1-40, or with bordeaux mixture, 4-4-50, as for brown rot.

Winter Injury: Trees so injured make a scant growth; many leaves turn yellow and fall about picking time; gum exudes at the crotches and about the trunk; sometimes the bark on the stock is entirely killed, in which case the tree languishes and finally dies.

Control: It is thought that heavy applications of highly nitrogenous fertilizers in late summer favor winter injury. Do not stimulate the tree to too active wood development. Cut out the gum pockets and cankers, and paint them with a heavy lead paint.

Chestnut

Bark Disease (Diaporthe parasitica): A fungous disease, attacking the bark of the American chestnut. Limbs and trunk are girdled, and the tree dies. The disease is present in many of the nurseries.

Control: Inspect nursery stock very carefully, especially about pruned stubs. Discard diseased trees. Make a careful examination of old trees, especially about old wounds and pruned stubs. If the disease is present, clean out the diseased wood with a gouge, and coat heavily with gas-tar. If the disease has progressed far, cut off diseased limbs or the whole tree and burn at once. Keep all wounds and pruned stubs covered with gas-tar.

Chrysanthemum

Leaf-spot (Septoria chrysanthemi): First appears as dark brown spots, which increase in size until the leaf dies.

Control: Pick and burn diseased leaves. Spray the plants with bordeaux mixture, 4-4-50.

Rust (Puccinia chrysanthemi): Reddish brown rust pustules on the leaves.

Control: Avoid wetting the foliage when watering. Spray as for Leaf Spot.

Corn

Ear-rot (Diplodia zeae): Several other organisms may cause an ear rot, but this is the more common one. The ear is imperfectly developed, soft, and overrun with a whitish mold. In many cases the husks and silk are also involved.

Control: Destroy old infected ears and stalks. Practice a rotation which will exclude corn for two years from or near the given plat of ground.

Rust (Puccinia maydis): Reddish pustules on the blades. Common on some varieties of sweet corn.

Control: No satisfactory method of control is known.

Smut (Ustilago zeae): Attacks stalks, ears, and tassels, producing abnormal boils or outgrowths. Will infect at actively growing points at any time.

Control: Rotate crops. Do not manure corn ground. Cut out smut and burn it. Soaking seed is of no avail.

Cotton

Anthracnose (Colletotrichum gossypii): Forms black or purplish colored spots on bolls. Disease also occurs on seed leaves and on the leaves and stems. Select seed from fields free from the disease. Rotate crops. Use disease-resistant varieties.

Root-rot (Ozonium omnivorum): Easily recognized by the sudden wilting and dying of the plants in the field.

Control: A combination of rotation of crops and deep fall plowing is effective.

Wilt (Fusarium vasinfecta): Causes a gradual wilt and eventual death of leaves and stems.

Control: Rotate crops. Secure seed of wilt-resistant varieties of cotton.

Cranberry

Blast or Scald (Guignardia vaccinii): The fungus causes a blast of the flowers and very young fruits, and attacks older fruits, causing them to appear scalded or watery.

Control: Spray five or six times with bordeaux mixture, 5-5-50, to which has been added 4 pounds of resin fish oil soap, making the first application just before the blossoms open. Long lines of hose are most satisfactory for this work, and the spraying must be done thoroughly.

Rot. (Acanthorhynchus vaccinii): A disease which cannot be distinguished from scald with the naked eye.

Control: Spray five or six times with bordeaux mixture, 5-5-50, to which has been added 4 pounds of resin fish oil soap, making the first application just before the blossoms open. Long lines of hose are most satisfactory for this work, and the spraying must be done thoroughly.

Hypertrophy (Exobasidium oxycocci): Appears on the young leaves soon after the water has been let off in the spring. The axillary leaf buds are attacked and produce short shoots with rather close, enlarged, swollen, and distorted leaves which are pink or light rose colour. The production of fruit is prevented or reduced.

Control: Early spraying with bordeaux mixture has been advised.

Cucumber

Blight or Mildew (Pseudoperonospora cubensis): A blighting and premature yellowing of the foliage.

Control: Spray with bordeaux mixture, 5-5-50. Commence to spray when the plants begin to run, and repeat every ten to fourteen days throughout the season.

Wilt (Bacillus tracheiphilus): This is a disease caused by bacteria that get into the sap tubes of the leaf and stem, clog and destroy them, causing the plant to wilt. The bacteria are distributed chiefly by the striped cucumber beetle.

Control: Control the striped beetle. Gather and destroy all wilted leaves and plants.

Wilt (caused by malnutrition): Excessive fertilizing with highly nitrogenous fertilizers will sometimes produce a peculiar curling and wilting of the leaves.

Currant

Anthracnose (Glaeosporium ribis): Small dark brown spots, chiefly on the upper surface of the leaf. The leaves finally turn yellow, and fall in July or August.

Control: Thorough applications of bordeaux mixture, 5-5-50.

Leaf-spot (Septoria ribis, Cercospora angulata etc.): Whitish spots with black centers. Appears in midsummer, and causes defoliation.

Control: Thorough applications of bordeaux mixture, 5-5-50.

Wilt or Cane-blight: A destructive fungous disease which causes the canes to die suddenly. Character of the wilting much like that produced by the cane-borer.

Control: No satisfactory method known. The most that can be done is to go over the patch three or four times during the summer, cut out and burn the blighted canes.

Ginseng. Blight (Alternaria panacis): Papery brown spots on the leaves, which spread until the whole leaf is involved. Also attacks the seed heads, producing a blast.

Control: In the spring before the plants come through the ground spray the soil thoroughly with copper sulfate, 1 pound to 10 gallons of water. As the plants are breaking through the soil, spray with bordeaux, 3–3–50. Spray repeatedly while the plants are coming through the ground, making a special effort to cover the stems. Keep all growth covered with spray throughout the summer. Spray the seed heads thoroughly just after the blossoms fall, and again when they are two-thirds grown. Destroy all diseased tops.

Fibre Rot (Thielavia basicola): Commonly called rust or rusty root, from the characteristic appearance. The plants eventually wilt and die.

Control: Treat the soil with acid phosphate at the rate of 1000 pounds to the acre. Dip the roots in bordeaux mixture, 3-3-50, before planting.

Mildew (Phytophthora cactorum): Attacks tops shortly after they come up.

Control: Spray with bordeaux mixture, 5–5–50. Commence to spray when the plants begin to run, and repeat every ten to fourteen days throughout the season.

Wilt (Acrostalagmus sp.): A sudden wilting of the whole plant, caused by the action of the fungus in the sap tubes of the root.

Control: Remove the wilted plants as soon as discovered in order to prevent further spread.

Root-rot: Caused by various soil organisms. Favored by wet, soggy soil.

Control: Underdrain the soil thoroughly. Golden-seal. - Consult treatments under Ginseng.

Gooseberry

Mildew (Sphaerotheca mors-uvce): A powdery mildew attacking the fruit and young growth of English varieties of gooseberpy.

Control: As soon as the leaves begin to unfold, spray with potassium sulfid, 1 ounce to 2 gallons of water.

Rust (AEcidium grossularice): Orange-colored rust pustules on the fruit and under side of the leaves.

Control: Spray with bordeaux mixture, 5–5–50. Commence to spray when the plants begin to run, and repeat every ten to fourteen days throughout the season. Keep down sedges and grasses.

Grape

Anthracnose (Sphaceloma ampelinum): Occurs on the fruit as a definite dark brown spot with a lighter auriole; on canes as deep pits with an elevated red margin, and on veins of the leaves, causing the leaves to crimp. Occurs on all varieties, especially Roger's hybrids. Not so abundant as formerly.

Control: It is said that an early spraying before the buds open with sulfate of iron, 100 pounds to 50 gallons of water, is very important.

Later sprayings for black rot will also be effective in preventing spread.

Black-rot (Guignardia bidwellii): The most serious disease of grapes east of the Rocky Mountains, especially southward. Attacks all green parts. Produces a brown circular spot on leaves, a black, elongated, sunken pit on petioles, canes, etc., and on the berry a brown rot with shriveling and wrinkling; finally the berry becomes black and hard.

Control: This disease may be controlled by timely applications of bordeaux mixture, 4–4–50. It is of great importance that spraying be done before rain storms, as the berry enlarges so rapidly.

Spray:

- When the third or fourth leaf has unfolded;
- As soon as the blossoms have fallen;
- When the berries are the size of a pea;
- In about two weeks. In a wet season make two more applications.

After July 20 make the bordeaux 4–2–50, or use ammoniacal copper carbonate. In case of dense foliage all applications except the first two should be made by hand. Attach trailers to the sprayer, and have two men following to apply the spray directly to the clusters. About ten acres can be sprayed in a day, and the total cost of labour and material should not exceed 75 cents per acre for each application.

California Vine-disease: An obscure disease, which destroyed thousands of acres of vines in California. Cause not known, and at present practically unknown and of no importance economically.

Crown-gall or Black Knot (Bacterium tumefaciens): A tu-merous, gnarled outgrowth on roots and stems, especially on European varieties. Frost injury often forms an infection court for the bacteria.

Control: Grub out and burn infected vines.

Downy Mildew or Leaf-blight (Plasmopara viticola): Appears in white frost like patches on under side of leaf, the upper side of the leaf showing a yellowish discoloration; gradually spreads to all parts of the leaf causing it to dry up.

Attacks the berry, which remains hard and white or gray. Worst on hybrids with vinifera blood; especially common on Delaware and Roger's hybrids. Widespread in North America.

Control: This disease may be controlled by timely applications of bordeaux mixture, 4–4–50. It is of great importance that spraying be done before rain storms, as the berry enlarges so rapidly.

Spray:

- When the third or fourth leaf has unfolded.
- As soon as the blossoms have fallen.
- When the berries are the size of a pea.
- In about two weeks. In a wet season make two more applications.

After July 20 make the bordeaux 4–2–50, or use ammoniacal copper carbonate. In case of dense foliage all applications except the first two should be made by hand. Attach trailers to the sprayer, and have two men following to apply the spray directly to the clusters. About ten acres can be sprayed in a day, and the total cost of labour and material should not exceed 75 cents per acre for each application.

Necrosis or Dead-arm Disease (Fusicoccum viticolum): Attacks shoots, and progresses from there to the old wood, causing a dry rot and eventual death of the vine.

Control: Inspect canes at trimming time, and use care not to leave those on which the brownish black spots are present. Train up renewals from the root, and cut off the old stem below the diseased area.

Treatment as for black-rot is efficacious.

Shelling or Rattles: Cause unknown. The berry breaks squarely off at its juncture with the pedicle. The leaves on such vines usually turn reddish brown about the margin. Powdery mildew is sometimes responsible for shelling.

Control: No method is known.

Anthracnose (Colletotrichum malvarum): Angular brown spots on leaves and stems which spread, killing the entire leaf.

Control: As for Rust.

Rust (Puccinia malvacearum): Attacks all parts of the plant, causing reddish brown pustules on affected parts; later leaving deep pits; may entirely destroy the leaves. It is abundant on the common mallow or "cheeses."

Control: Eradicate the mallow; pick off diseased leaves in the fall, and burn all litter. Repeat in the spring, and spray new growth thoroughly with bordeaux mixture, 4-3-50. Spray every week until the flower-stalks are well developed. Lettuce.

Leaf Perforation (Marssonia perforans): Dead areas in the leaves which finally drop out. Also on veins of the leaves.

Downy Mildew (Bremia lactucce): Yellow spots on the upper surface of the leaf, accompanied by a frosty growth on the opposite side.

Control: Destroy infected plants. Keep water from the leaves; furnish water by means of subirrigation.

Drop or Rot (Sclerotinia libertiana): Base of the leaves or stem rots off, allowing leaves to drop.

Control: Sterilize the soil with steam before planting.

Rosette (Rhizoctonia sp.): A rotting or damping-off of the stem. Late affected plants have a rosetted appearance.

Control: Start seed in steam-sterilized soil, and transfer to beds that have been sterilized with steam, as for Drop.

Muskmelon

Anthracnose (Colletotrichum lagenarium): Dead spots on the leaves and stems and sunken pits on the fruit. Thorough and frequent spraying with bordeaux mixture will hold this disease in check.

Downy Mildew: The same disease as on cucumbers. Often very destructive.

Control: A satisfactory method is not known. Spraying as for cucumber mildew has not proved effective.

Nursery Stock: Foliage on young trees is apt to be attacked by various leaf-spot fungi. The damage comes in reducing growth, thus often making seconds. Several applications of bordeaux mixture to keep the new growth protected are beneficial.

Oats

Rust (Puccinia coronata): A red rust of the blades.

Control: There is no known method of control.

Onion

Mildew (Peronospora schleideniana): Causes a wilt or blight of the leaves.

Control: Spray with bordeaux mixture, 5–5–50, to which has been added one gallon of resin-sal-soda sticker. The first application should be made when the third leaf has developed, and the application should be repeated every ten days until the crop is harvested.

Smut (Urocystis cepulae): Forms black pustules on the leaves and bulbs. Seedlings may be killed outright.

Control: Onions from sets or from seed started in soil free from the disease seldom have the smut. Practice crop rotation. Drill into the soil with the seed 100 pounds of sulfur and 50 pounds of air-slaked lime to the acre.

Pea. Mildew (Erysiphe polygoni): A powdery mildew on pods and leaves.

Control: Dust dry sulfur over the plants, repeating the operation if necessary.

Pod Spot and Leaf-spot (Ascochyta pisi): Black circular spots on stems, leaves, and buds.

The fungus grows through the pod into the seed, and is thus carried through the winter.

Control: Select pods free from spots, and save the seed from these for the next year's planting. On a large scale have a clean seed garden in which to grow clean seed for the following year.

Peach. Blight (Coryneum beyerinkii): A spotting, gumming and death of the buds and twigs, particularly in the lower part of the tree. The fruit drops. Especially serious in California.

Control: For California conditions two applications of spray are made:

- In November or December
- In February or March.

This also controls leaf-curl. Bordeaux mixture, 5–5–50, or lime-sulfur, 1-10, may be used.

Brown-rot (Scerotinia fructigena): Causes a rot of the fruit, and often runs down the spur, forming a canker in the limb. Also produces brownish irregular spots on the leaves.

Control: Spray with self-boiled lime-sulfur, 8-8-50, adding 2 pounds of arsenate of lead. Spray first about time shucks are shedding from young fruit; second, two to three weeks later, and third, about one month before the fruit ripens. Omit the arsenate of lead from the third spraying.

On early maturing varieties two applications may be sufficient. Spraying within a month of picking time is apt to leave the fruit spotted. It is especially important that sprayings be made before a continued storm period. Destroy rotten peaches. The rotten ones on the ground are as great a menace (especially if plowed under) as. those on the tree, as the fungus winters readily on the fallen mummies. Brown-rot also occurs on cherries, plums, apricots, and sometimes on apples and pears.

Leaf-curl (Exoascus deformans): Causes the leaves to crimp and curl and often to turn bright red. Also causes shoots to swell and become distorted.

Control: In an infected orchard more than 90 per cent of the curl can be controlled the first year. The second year control should be complete. The secret of control of leaf-curl lies largely in the thoroughness with which the work is done. A number of spray substances may be used. A single thorough application before the buds swell in the spring is sufficient. Every bud must be covered and from all sides. Lime-sulfur as applied for San Jose scale will control curl. Commercial lime-sulfur, 1-20; bordeaux mixture, 4–4–50; or a simple solution of blue vitriol in water, 2–50, are all specifics.

Little-peach: A disease that in its early stages resembles yellows. It differs from yellows in producing small fruit that matures later than normally. Fruit does not have the small red spots characteristic of yellows, nor are there slender sickly branches. The cause of this disease is unknown. Apparently spreads more rapidly than yellows and commonly destroys the affected tree sooner.

Powdery mildew (Sphcerotheca pannosa): A whitish powdery growth on the young shoots and leaves, and whitish spots on the fruit.

Control: Self-boiled lime-sulfur as for Rot.

Root-gall, Root-Knot, Crown-gall, Hairy-root (Bacterium tumefaciens): Hairy roots or tumerous outgrowths on the roots and root crowns; sometimes occurs on trunks and limbs. Primarily a nursery disease. Does not seem to be a serious disease on peaches in the North, but is reported as very serious in the South. Attacks a wide range of orchard plants, including apple, pear, brambles, grape, etc.

Control: Reject all stock showing symptoms.

Rosette. – An obscure southern disease of peach trees and some kinds of plums, characterized by bunchy growths containing very many rolled and yellowish leaves which fall prematurely. The tree dies the first or second year. There is no premature fruit as in yellows. It is often accompanied by gummosis of the roots. The disease is communicable by budding, and it may enter through the roots.

Scab or Black-spot (Cladosporium carpophilum): Black scablike spots on the fruit, often causing it to crack deeply.

Control: Self-boiled lime-sulfur, as applied for Brown-rot.

Yellows: A fatal disease of peaches; also attacks nectarine, almond, apricot, and Japanese plum. Cause unknown. The first symptom in bearing trees is usually the premature ripening of the fruit. This fruit contains definite small red spots, which extend towards the pit. The second stage is usually the appearance of " tips," or short, late, second growths upon the ends of healthy twigs, and which are marked by small, horizontal, usually yellowish leaves.

The next stage is indicated by very slender shoots, which branch the first year and which start in tufts from the old limbs, bearing narrow and small yellowish leaves. Later the entire foliage becomes smaller and yellow. In three to six years the tree dies. The disease spreads from tree to tree. It attacks trees of any age.

Preventive: Pull up and burn all trees as soon as the disease appears. Trees may be reset in the places from which the 1 yellows " trees were taken. Laws aiming to suppress the disease have been enacted in most peach-growing states, and the enforcement of them will keep the disease well under control.

Pear

Blight (Boxillus amylovorus): A very serious bacterial disease. Bacteria winter just at the edge of the dead wood in trees blighted the previous year. With the advent of warm spring days they ooze through the bark in sticky drops and are carried by bees and flies to blossoms. The blossoms blight, and the spur may also blight. Plant-lice carry bacteria from blighted blossoms to spurs and shoots. If a spur becomes blighted, the bacteria may spread in the bark of the limb, causing a depression or canker.

This may girdle the limb and cause its death. The leaves turn black and stick tenaciously, even through the winter. Succulent water sprouts are very apt to blight and cause large cankers. Generally distributed in North America, and known only in America. Attacks apple, quince, mountain ash, hawthorn; the Spitzenburgh is specially liable to attack.

Control: Clean up hedgerows of hawthorn, old blighted pear trees and apple trees. In early spring cut out the blight of the previous year and disinfect the stubs with corrosive sublimate, 1-1000. Clean out cankers with a sharp knife, and disinfect. Paint over with lead paint. At blossoming time make a systematic daily inspection for blossom blight, and break it out. Watch for blight in the shoots.

When it appears get a long-handled pruning-hook, fasten a sponge near the knife, and saturate it with corrosive sublimate solution, 1-1000. Clip out the blighted twigs, cutting five or six inches below the blight, and sop the pruned stub with the sponge. During a blight epidemic, drop all other work. The work must be done systematically and persistently, or not at all. One week's work may save the pear crop and the pear trees.

Leaf-blight and Cracking of fruit (Fabrea maculata): Attacks nursery stock of pears and quinces, beginning as small circular brown spots on the leaves. These spread, and if numerous cause the leaf to fall. The same disease produces a black spot or pit on the fruit.

Control: In nurseries spray with bordeaux mixture, 4-4-50. Owing to the nearness of the fungus (on the twigs) and the slowness with which the pear-leaf unfolds, two applications of spray before the blossoms open are sometimes necessary, and one immediately after they fall.

Use lime-sulfur, 1-50, or bordeaux, 3-3-50, with perhaps one additional application.

Leaf-spot (Mycosphaerella sentina): Small lecticular spots with white centers on leaves. Spots become so numerous as to cause defoliation. The fungus is known only on leaves, and it winters on them.

Control: Burn fallen leaves. Owing to the nearness of the fungus (on the twigs) and the slowness with which the pear-leaf unfolds, two applications of spray before the blossoms open are sometimes necessary, and one immediately after they fall. Use lime-sulfur, 1-50, or bordeaux, 3-3-50.

Scab (Venturia pyrina): Greenish brown or black spots on leaves and fruit, arresting growth and often causing fruit to crack. Severe on Flemish Beauty. Often attacks pedicles of fruits and causes them to drop, and may even cause defoliation. Is different from apple scab, but behaves much like it. Differs especially in the fact that the fungus winters on the twigs as well as on fallen leaves.

Control: Owing to the nearness of the fungus (on the twigs) and the slowness with which the pear-leaf unfolds, two applications of spray before the blossoms open are sometimes necessary, and one immediately after they fall. Use lime-sulfur, 1-50, or bordeaux, 3-3-50.

Rust (Gymnosporangium globosum): Having the same habits and appearance as apple rust.

Control: Owing to the nearness of the fungus (on the twigs) and the slowness with which the pear-leaf unfolds, two applications of spray before the blossoms open are sometimes necessary, and one immediately after they fall. Use lime-sulfur, 1-50, or bordeaux, 3-3-50.

Plum

Black-knot (Plowrightia morbosa): A black tumerous swelling from one to several inches in length, appearing on the limbs and twigs of American plums and sour cherries. Point of attack is usually under a bud or in crotches. Confined to America. A very serious disease. In some regions it has destroyed the plum industry. It was once supposed to be caused by an insect.

Control: Burn all affected parts in the fall. Cut several inches below the swelling. A badly infected tree should be cut down at once, as there is no hope of saving it. Many states have a law requiring the destruction of affected trees.

Leaf-rust (Puccinia pruni-spinosce): Small circular powdery spots of yellowish brown on the under surface of the leaves, and reddish spots on the upper surface directly above them.

Control: Early spraying with bordeaux, 3-3-50, or self-boiled lime-sulfur, 8-8-50.

Potato

Early blight (Alternaria solani): A blight of foliage beginning as an even circular spot and coming early in the season, usually in July. Progresses slowly. This disease does not attack the tubers.

Control: Bordeaux mixture at intervals of ten days, beginning when plants are 6-8 in. high.

Late blight and Potato-rot (Phytophthorainfestans): The fungus winters in the tuber, which shows a faint pinkish tinge and a dry rot. Diseased tubers are planted, the fungus fruits on the cut surface and its swarm spores pass through the soil-water to the leaves which touch or are buried in the soil. An extensive irregular blighted area covers the leaf, the under surface of which may have a mildewy appearance. The disease spreads very rapidly. Later spores are washed down to the tubers and infect them.

Control: Can be controlled successfully by the use of bordeaux mixture, 5-5-50. It is always profitable to spray at least three times, and in a wet season six or more applications shouid be made. As the vines increase in size, greater quantities of spray and more nozzles must be used. Use from 40 to 100 gallons of spray mixture per acre.

Dry-Rot and Wilt (Fusarium oxysporum): A dry rot of the tuber in storage and wilt of plants in the field. Can be detected in the seed tuber before there is any external appearance by examining a section near the stem end.

A black ring or chain of dots near the surface is indicative of the rot. Infection frequently takes place through wounds.

Control: Reject all diseased tubers for seed. Practice a rotation in which potatoes are not grown on the soil for at least two years.

Scab (Oospora scabies): A scabby and pitted roughness of potato tubers. Lime, ashes or manure added to the soil increases the amount of scab by favoring the growth of the fungus. It has become one of the serious diseases of the potato.

Control: Do not plant on land which has grown scabby potatoes. Plant clean seed. If only scabby seed is at hand, soak the uncut tubers in a solution of formalin, 1 pint in 30 gallons of water, for two hours. Drain, cut, and plant in clean soil. Use the formalin solution over and over. The same fungus also attacks beets.

Pumpkin

Quince. Black-rot (Sphaeropsis malorum): A trouble which usually appears at the blossom end of young quince fruits, causing them to become black and hard, with a dry rot of the tissue.

Rust: The organism causing this disease is of the same habit and nature as that causing apple rust.

Radish

White rust or Mildew (Albugo candidus): A whitish powdery growth on the leaves and petioles, often causing distortion.

Control: Steam-sterilize the soil before planting.

Raspberry

Anthracnose (Glaeosporium venetum): Circular or elliptical, gray scab-like spots on the canes.

Control: Avoid taking young plants from diseased plantations. Remove all diseased canes as soon as the fruit is picked. Practice frequent rotation.

Crown-gall or Root-call (Bacterium tumefaciens): Tumerous outgrowths on the roots, especially on red varieties. It is contagious and destructive.

Control: Never set plants which have galls on the roots. Avoid setting on infested land.

Red or Orange rust (Gymnoconia interstitialis): A dense red powdery growth on the under side of the leaves of black varieties and of blackberries.

The fungus hibernates in the roots.

Control: Dig up and destroy infected plants.

Rice. Blast, Blight or Rotten-neck (Piricularia oryzae): An extensive paling and drying of leaf and stem, and a partial failure of the heads to fill.

Control: The selection of early maturing varieties is advisable. Burn stubble and trash left in the fields.

Rose

Black leaf-spot (Actinonema rosae): Attacks the full-grown leaves, first appearing as small black spots, but later covering nearly or quite the whole surface with blotches. The spots have frayed edges.

Control: Spray with ammoniacal copper carbonate, beginning with the first appearance of the spots and continuing at intervals of one week until under subjection.

Mildew (Sphaerotheca pannosa): A white powdery mildew on the new growth.

Control: For greenhouse roses keep the steam pipes painted with a paste made of equal parts lime and sulfur mixed with water. Out-of-door roses should be dusted with sulfur flour or sprayed with potassium sulfid, 1 ounce to 3 gallons of water. Spinach.-

There are numerous fungous diseases of this crop, but a

practical method of control has not been developed. The best that can be done is to rotate crops.

Strawberry

Leaf-spot or leaf-blight (Mycosphaerella fragarice): Small purple or red spots appearing on the leaves. They increase in size and make the leaf appear blotched. The fungus passes the winter in the old diseased leaves that fall to the ground.

Control: In setting new plantations remove all diseased leaves from the plants before they are taken to the field. Soon after growth begins, spray the plants with bordeaux mixture, 4-4-50. Make three or four additional sprayings during the season. The following spring spray just before blossoming, and again in ten to fourteen days. If the bed is to be fruited again, mow the plants and burn over the bed as soon as the crop is off.

Mildew (Sphaerotheca castagnei): A whitish cobweb-like mildew on fruit and leaves, causing the latter to curl.

Control: Spraying as for leaf-spot; dusting with sulfur flour.

Sweet-Potato

Black-rot (Ceratocystis fimbriata): Causing black shank of the plant and a black rot of the tuber. The spots on the tuber are greenish black, from a quarter of an inch to four inches in diameter and extending for some distance into the tissue.

Control: Never use affected potatoes from which to grow sprouts. Steam-sterilize the soil in the hotbed. Practice rotation.

Rots: The sweet-potato is susceptible to a large number of rots, soft, dry, hard, white, etc. In practically all cases the organism producing the disease is an inhabitant of the soil. The best method of preventing these diseases is to use perfectly sound potatoes for sprouts and plant on soil which has not grown sweet-potatoes for several years.

Tobacco

Root-rot (Thielavia basicola): A rot of the main root and dwarfs the plants. Occurs both in seed-bed and field.

Control: Steam-sterilize the seed-bed by the inverted pan method. Rotate crops. Avoid liming, and add acid fertilizers.

Wilt (Bacterium solanacearum): A wilt of the plants caused by bacteria.

Control: Very difficult to control, as the organism lives in the soil for years. Never plant on land known to be diseased. Do not cultivate related plants, as potato, tomato, egg plant, or pepper, on the same soil. Transplant early, and avoid breaking the roots. Where tobacco is grown under shade (as is now a common practice) the soil should be steam-sterilized.

Tomato

Bacterial Wilt (Bacterium solanacearum): A wilt of the plants caused by bacteria.

Control: Very difficult to control, as the organism lives in the soil for years. Never plant on land known to be diseased. Do not cultivate related plants, as potato, tomato, egg plant, or pepper, on the same soil. Transplant early, and avoid breaking the roots. Where tobacco is grown under shade (as is now a common practice) the soil should be steam-sterilized.

Blight or Scab (Cladosporium fulvum): Soft brown irregular spots on the under surface of the leaves. The upper surface becomes spotted with yellow. The leaves finally wither and die. Most serious in the greenhouse.

Control: In mild cases the disease can be prevented by picking off the affected leaves. In severe cases spray with bordeaux mixture, 4-4-50, at intervals of ten days.

Downy mildew (Phytophthora infestans): The same fungus that causes Potato-blight.

End-rot: Not well understood, and no method of control is known.

Leaf-spot (Septoria lycopersica): A serious disease attacking leaves and stems. At first small spots appear, which spread until the whole leaf is consumed. (In severe cases the fruit may also be attacked.

Control: Spray with bordeaux mixture, 4-4-50, making the first application two weeks after the plants are set out, and repeating every two weeks throughout the growing season.

Control: A diseased condition of forced tomatoes characterized by rolled or curled leaves, distended veins, and by swollen areas having a frosty appearance on leaf veins, petioles and stem. This condition may be brought about by insufficient light, too much water in soil, excessive fertilization, high soil temperature.

Prevention: Avoid conditions favorable for the disease. Provide good ventilation in forcing-house; in field, cultivate deep and avoid topping plants.

Violet

Leaf-spots and leaf-blights: A number of different organisms are responsible. Usually not very destructive.

Control: Destroy affected plants; use fresh soil for new plantings; spray the foliage in the summer and fall with bordeaux mixture, 4-4-50.

Root-rot (Thielavia basicola): The same as the root-rot of tobacco. The plants make poor growth, owing to the fungus on the roots.

Control: Start in steam-sterilized soil, and transfer to sterilized beds.

Chapter 3

Majority of Plant Pathology

The reliance of humans on plants for food, fibres and other resources, understanding plant diseases and their control is of vital importance to our survival. Plant disease epidemics can cause famines, eliminate a thriving industry or poison animals and humans. Plant diseases are the result of infection by other organisms that adversely affect the growth, physiological functioning and productivity of a plant, manifesting outwardly as visible symptoms. Parasitic organisms that cause disease are called pathogens.

The vast majority of plant pathogens are fungi, however, plant diseases are also caused by insects, bacteria, nematodes, viruses and phytoplasmas. Disease-like symptoms can also be caused by abiotic factors, such as temperature, light, chemical agents and water or nutrient deficiencies.

Symptoms of disease include death and destruction of host tissue, wilting, abnormal growth and differentiation and discolouration of host tissue. Some parasites, called necrotrophs, secrete enzymes that kill host tissue, extract nutrients from the cells and then live in the dead tissue. The necrotic lesions caused by pathogens can be localised or extensive. Local necrotic lesions appear as discrete necrotic areas, while extensive, or spreading lesions spread until the whole organ or plant is killed.

Wilting occurs when water loss is greater than water intake. It results from either: interference with water and nutrient absorption at the roots, interference with water conduction within the plant (i.e. infection of the vascular tissue), or loss of control of transpiration. Abnormal growth and differentiation results from deviation from the complex balance of interrelated reactions that take place in plants. Parasites can alter the hormonal balance in plants causing an abnormal increase in the size or number of cells, resulting in abnormal growth and differentiation, for example, the formation of galls.

Discolouration of tissue is most commonly by chlorosis or mosaics of leaves, both of which can have a number of causes. Anything that interferes with the production of chlorophyll causes leaves to turn yellow, or chlorotic. Mosaicism is a symptom of many virus infections and is characterised by alternating light and dark green areas on the leaves.

Table. Some of the Most Common Terms used to Describe Symptoms of Plant Diseases.

Blight	A disease characterised by widespread death of plant tissue.
Canker	A sunken necrotic lesion often of a main stem, branch or root.
Damping-off	Collapse and rot of seedlings near soil level before emergence or soon after emergence caused by *Pythium* spp., *Phytophthora* spp., *Fusarium* spp., and *Rhizoctonia* spp.
Dieback	Partial defoliation, twig and branch death and even complete death of plants.
Downy mildew	White or grey 'bloom' on leaves and stems caused by production of sporangiophores and sporangia by members of the Peronosporales (downy mildew fungi).
Gall	An abnormal growth or swelling produced as a result of pathogenic invasion.
Mosaic	Patchy variation of normal green colour in leaves, usually light and dark green

	mosaic, symptomatic of many viral diseases.
Powdery mildew	White powdery 'bloom' on the plant surface caused by the production of fungal mycelium, conidiophores and conidia by members of the Erysiphales (powdery mildew fungi).
Pustule	A blister-like spore mass breaking through a plant epidermis.
Rot	Disintegration of tissue, often caused by enzymes or toxins produced by pathogens.
Rust	Rust-coloured pustules formed by members of the Uredinales (rust fungi).
Scab	A discrete, superficial roughened lesion.
Smut	A disease characterised by black spore masses on leaves, stems or inflorescences, caused by members of the Ustilaginales (smut fungi).
Vascular wilt	A disease in which the pathogen is confined to the vascular system of the host and in which wilting is a characteristic symptom; plants lose their turgidity and become flaccid, leaves collapse.

PATHOGEN SURVIVAL AND DISPERSAL OF PLANT PARASITES

The survival of a parasite between cropping seasons and its effective dispersal to uninfected plants are crucial aspects of the plant disease cycle. If either of these is prevented, the disease will not occur. Most pathogens possess mechanisms to survive intercrop periods or periods of unfavourable environmental conditions. The spread of inoculum can be airborne, soil-borne, water-borne, seed- or clone-borne, or vector-borne. Airborne inoculum can travel for great distances, even across oceans, while soil-borne inoculum is rarely spread any great distance. Many pathogens are dispersed by more than one mechanism.

SURVIVAL

Continuous Infection Chains

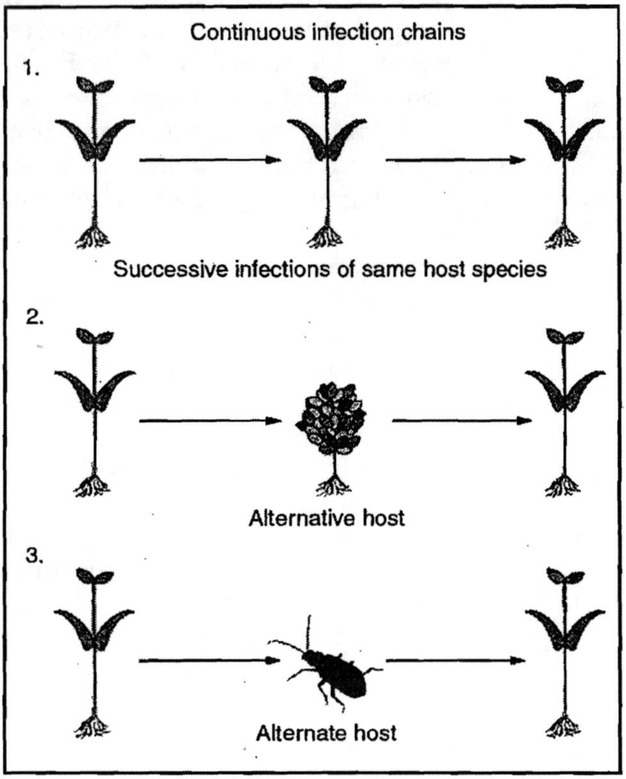

Fig. Continuous Infection Chains

The above diagram shows 3 examples of continuous infection chains.

1. The pathogen can repeatedly infect plants of the same host species.
2. Alternately infect the main crop species and another related species.
3. Infect a plant species via a vector, such as an insect, which transmits the pathogen from an infected plant to another plant of the same species.

The survival of most plant pathogens requires the

repeated infection of host plants. This is known as the infection chain. The infection chain can be continuous, or discontinuous (incorporating a resting phase).

Continuous infection chains can involve the same or alternative hosts. The parasite survives by continually infecting plants of the same host species, or, if the host species has a dormant or intercrop period, alternately infecting the main crop species and another, often related, species, the alternative host. If the alternative host does not display disease symptoms, it is called a disease carrier.

The parasite does not form resting structures, and it is dependent on the presence of a susceptible host species. If the crop species grows throughout the year, for example, in tropical areas, the parasite can survive by continuously infecting new individuals of the same host species. Where the crop is not grown year-round, self-sown individuals of crop species that grow by the side of the road or as weeds in paddocks can act as hosts for parasites between cropping seasons.

There are some plant pathogens that cannot be transferred directly from one plant to another plant of the same species. They require another, completely unrelated, species to act as a vector. These vectors are usually, but not always, insects, and are referred to as alternate hosts. Unlike the case of alternative hosts, which are used opportunistically and where necessary, the alternate host is a necessary step in the infection cycle. If there is no alternate host available, even if there are susceptible plants available, the infection chain is broken.

Discontinuous Infection Chains

The above diagram illustrates 2 examples of discontinuous infection chains. The pathogen survives on the surface of its host plant (epiphytic phase) or it survives in the soil during the intercrop period (saprophytc phase) and is therefore able to infect the new crop of the host plant species.

Discontinuous infection chains usually involve an epiphytic, saprophytic or resting phase. During an epiphytic phase, the pathogen survives on the surface of their host or other plants in a non-parasitic relationship as an epiphyte. Pathogens that go through a saprophytic phase survive during intercrop periods on diseased plant debris or other organic matter on or in the soil.

Some of them can compete very effectively with the normal soil microflora. Others specifically inhabit the diseased plant debris.

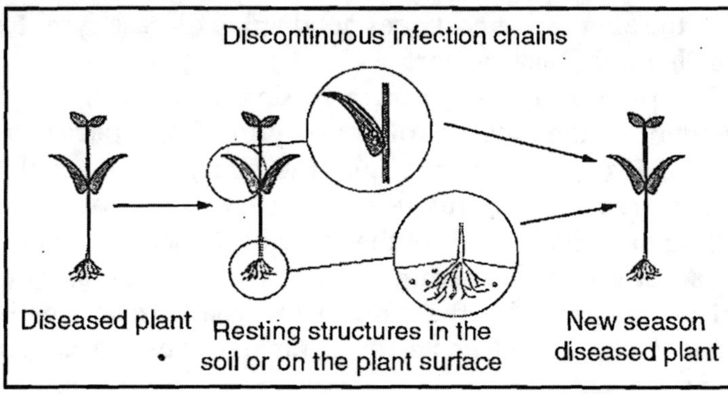

Fig. Discontinuous Infection Chains

Fungi and nematodes are able to form resting structures that enable them to survive long periods without a suitable host, or when environmental conditions are unfavourable. The resting spores (*oospores, teliospores* or *chlamydospores*) of some fungi can survive for twenty years or more. Some of them are triggered to germinate only by secretions from the roots of suitable plants, reducing the risk of germinating without an available host.

Other fungi produce sclerotia, which can also survive in the soil for periods ranging from months to years. Fungi can also produce sexual fruiting structures (such as *cleistothecia, perithecia* and *pseudothecia*) during the resting stage. Some fungi go through a resting stage after infection, called a latent infection.

For example, loose smut fungus of wheat infects wheat embryos in the flowers, becomes dormant and is activated again when the seed germinates. When the plant matures, the fungus produces teliospores in the place of inflorescences.

Nematode eggs can survive for long periods in egg cysts or gelatinous egg masses, which reduce the rate of egg desiccation.

DISPERSAL OF INOCULUM

Inoculum can be classified as primary or secondary inoculum. Primary inoculum consists of propagules of a pathogen that start the disease cycle in a new growing season. Secondary inoculum distributes the pathogen within the main growing season of the crop. It is usually secondary inoculum that leads to the development of epidemics.

Inoculum can be carried from plant to plant by air currents, through the soil, by water splash, or via a vector species, such as an insect, other animal, fungus or plant. Insect vectors often have piercing and sucking mouthparts, penetrating the plant surface and providing an infection pathway for viruses, phytoplasmas and plant parasitic protozoa that they are carrying inside their bodies.

Some bacteria and fungi are spread by sticking to the outside of insect vectors. Disease can also be dispersed in planting material, such as clones and seeds. Seed-borne inoculum can be mixed in with the seed during harvesting, attached to the surface of the seed, or present inside the seed, having already infected the seed or embryo. About one fifth of known plant viruses are dispersed via seed.

DISTRIBUTION OF DISEASED PLANTS

The distribution of diseased plants within a population can provide information about the source of inoculum or the nature of the vector. Random distribution of diseased plants is not common. It can be a result of seed-borne infection, insect-borne inoculum or airborne inoculum being introduced from a long way away. Aggregations of diseased plants are more common, indicating a random distribution of inoculum, which has then spread from the originally infected plants.

A disease dispersed by aphids, or a root-infecting fungus might be expected to produce a pattern of aggregation as the inoculum spreads from the initial diseased plants. Formulae and computer programs exist that can analyse the pattern of disease spread, and help to locate the source or the vector of the inoculum

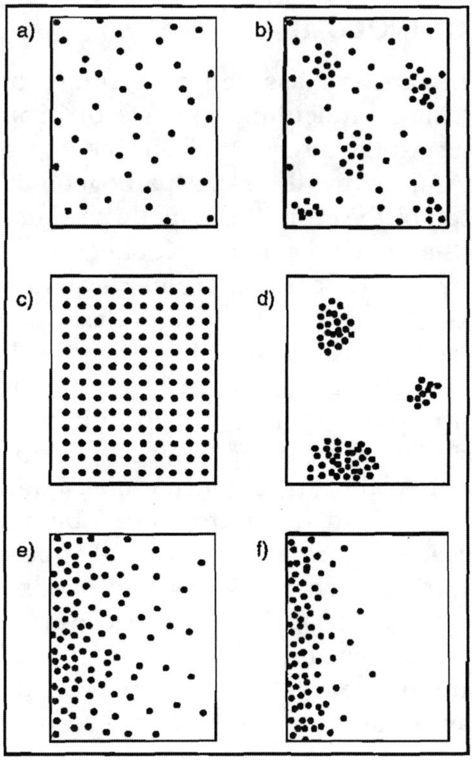

Fig. Different Types of Distributions of Diseased Plants

This diagram shows different types of distribution of diseased plants that may be found within a population: (a) Random distribution; (b) Aggregations of diseased plants; (c) Regular distribution; (d) Patch distribution; (e) and (f) Disribution in the form of a gradient, with (f) showing a steeper gradient than (e).

Regular distribution of diseased plants is highly unusual in the field. However, it could occur in vegetatively propagated crops if all planting material was infected, or if a previous crop that was evenly spaced left pathogens in the soil that were able to infect the next crop. Patch distribution of disease is characteristic of soil-borne diseases, such as root-rotting fungi or nematodes, or diseases carried by soil-inhabiting vectors. Soil-inhabiting organisms usually spread very slowly, as do the diseases carried by these vectors, hence the patchy distribution.

A very common pattern of disease distribution is a gradient. Gradients usually indicate that the source of the inoculum is outside the crop, and the steepness of the gradient is proportional to the closeness of the source.

Other factors can influence the slope of the gradient, such as how the disease is spread (eg. Crawling insects v flying insects). Vector movement has a much greater influence on the spread of disease than vector number. A few active vectors will spread the disease much more rapidly that many static vectors.

INFECTION PROCESS

The "infection process" can be divided into three phases: pre-entry, entry and colonisation. It encompasses the germination or multiplication of an infective propagule in or on a potential host through to the establishment of a parasitic relationship between the pathogen and the host. The process of infection is influenced by properties of the pathogen, the host and the external environment. If any of the stages of the infection process is inhibited by any of these factors, the pathogen will not cause disease in the host.

While some parasites colonise the outside of the plant (ectoparasites), pathogens may also enter the host plant by penetration, through a natural opening (like a stomatal pore) or via a wound. The symptoms of the diseases produced by these pathogens result from the disruption of respiration, photosynthesis, translocation of nutrients, transpiration, and other aspects of growth and development.

PRE-ENTRY

Before a pathogen can penetrate a host tissue, a spore must germinate and grow on the surface of the plant. In the case of motile pathogens, they must find the host and negotiate its surface before entering the host. Some pathogens develop specialised penetration structures, such as appressoria, while others utilise pre-existing openings in the plant's surface, such as wounds or stomatal pores. Plant viruses are often transported and introduced into the plant via vectors such as fungi or insects.

The initial contact between infective propagules of a parasite and a potential host plant is called inoculation. Pathogens use a variety of stimuli to identify a suitable entry point. Several fungi use topographical cues on the plant surface to guide them towards a likely stomatal site.

Once the hypha reaches a stoma, volatile compounds escaping from the pore appear to provide a signal for the formation of a specialised penetration structure, the appressorium. Sugars, amino acids and minerals secreted by plants at the leaf surface can non-specifically trigger spore germination or provide nutrition for the pathogen. Some pathogenic spores will not germinate in the absence of these substances.

Pathogen development is influenced by temperature, moisture, light, aeration, nutrient availability and pH. The conditions necessary for survival and successful infection differ between pathogens.

ENTRY

Pathogens exploit every possible pathway to enter their host, although individual species of pathogen tend to have a preferred method. Fungal pathogens often use direct penetration of the plant surface to enter the host. This requires adhesion to the plant surface, followed by the application of *pressure* and then *enzymatic degradation* of the cuticle and cell wall, in order to overcome the physical barriers presented by the plant's surface. During the degradation of the cuticle and wall, a succession of genes are switched on and off in the pathogen, so that cutinase, followed by cellulase, then pectinase and protease are produced, attacking the cuticle, cell wall, and middle lamella in the order that they are encountered.

The pressure needed for the hypha to penetrate the cell

wall is achieved by first firmly attaching the appressorium to the plant surface with a proteinaceous glue. The cell wall of the apressorium then becomes impregnated with melanin, making it watertight, and capable of containing the high turgor pressure that builds up within the appressorium.

The point of the appresorium that is in contact with the cuticle is called the penetration pore, and the wall is thinnest at this point. The increasing turgor pressure causes the pore to herniate, forming a penetration peg, which applies huge pressure to the host cuticle and cell wall.

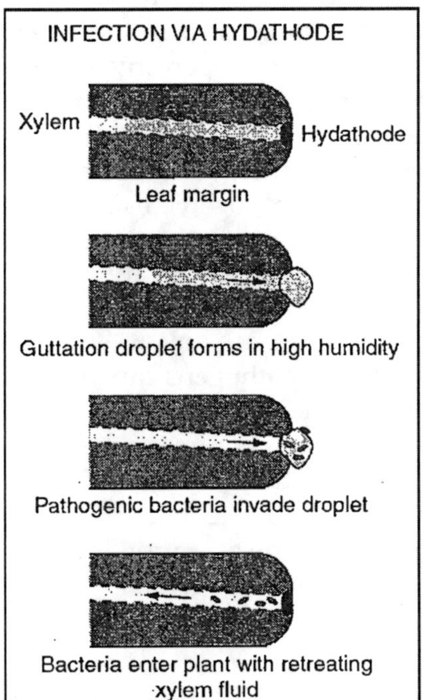

Fig. Pathogen Invasion Via an Hydathode

This diagram illustrates the way in which a pathogen invades a host plant via an hydathode. In conditions of high humidity, a guttation droplet forms at the hydathode and if the plant is exposed to pathogenic bacteria, the water droplet may be invaded. When humidity decreases the droplet is drawn back into the plant, carrying the bacteria within it.

The alternative pathway for pathogen entry is via a pre-existing opening in the plant surface. This can be a natural opening or a wound. Pathogenic bacteria and nematodes often enter through stomatal pores when there is a film of moisture on the leaf surface.

Fungi can also penetrate open stomata without the formation of any specialised structures. Some fungi form a swollen appressorium over the stomatal aperture and a fine penetration hypha enters the airspace inside the leaf, where it forms a sub-stomatal vesicle, from which infection hyphae emerge and form haustoria in surrounding cells.

Also vulnerable to pathogen invasion are hydathodes, pores at the leaf margin that are continuous with the xylem. Under particularly humid conditions, droplets of xylem fluid (guttation droplets) can emerge at the surface of the leaf where they can be exposed to pathogenic bacteria, which then enter the plant when the droplet retreats back into the hydathode as the humidity decreases. Lenticels are raised pores that allow gas exchange across the bark of woody plants. They exclude most pathogens, but some are able to enter the plant via this route. Some specialised pathogens can also use more unusual openings, such as nectaries, styles and ectodesmata. Entry through a wound does not require the formation of specialised structures, and many of the pathogens that utilise wounds to enter the plant are unable to penetrate the plant surface otherwise. Most plant viruses entrer through wounds, such as those made by their insect vectors.

COLONISATION

A successful infection requires the establishment of a parasitic relationship between the pathogen and the host, once the host has gained entry to the plant. There are two broad categories of pathogens are biotrophs (those that establish an infection in living tissue) and necrotrophs (those that kill cells before colonising them, by secreting toxins that diffuse ahead of the advancing pathogen). These two kinds of pathogens are also sometimes known as 'sneaks' and 'thugs', because of the tactics they use to acquire nutrients from their hosts.

The toxins produced by necrotrophs can be specific to the host or non-specific. Non-specific toxins are involved in a broad range of plant-fungus or plant-bacterial interactions, and will therefore not usually determine the host range of the pathogen producing them. Necrotrophs often enter the plant through wounds and cause immediate and severe symptoms. An intermediate category of parasite is the hemibiotrophs, which start off as biotrophs and eventually become necrotrophic, employing tactics from both classes of pathogen.

Pathogens that colonise the surface of plants, extracting nutrients through haustoria in epidermal or mesophyll cells are termed ectoparasites. The haustoria are the only structures that penetrate the host cells. Some parasites colonise the area between the cuticle and the outer wall of the epidermal cells, penetrating host epidermal and mesophyll cells with haustoria. These are called sub-cuticular infections. Pathogens can also form colonies deeper in the plant tissues. These are mesophyll and parenchyma infections, and can be necrotrophic, hemibiotrophicor biotrophic relationships.

Necrotrophs do not produce specialised penetration structures. Instead, they kill host cells by secreting toxins, then degrade the cell wall and middle lamella, allowing their hyphae to penetrate the plant cell walls and the cells themselves. In hemibiotrophic infections, intercellular hyphae can form haustoria in living mesophyll cells, but as the lesion expands under favourable conditions, those heavily parasitised cells at the inner, older part of the colony collapse and die.

A similar sequence of events can take place in plants infected by burrowing nematodes. Viruses, mildews and rusts develop specialised biotrophic relationships with their hosts. Intercellular hyphae of downy mildew colonise host mesophyll cells and form haustoria.

The mildew sporulates and the infected cells eventually die, although necrosis is delayed and contained, compared to that caused by necrotrophic pathogens. Rust fungi can also delay senescence in infected cells while they sporulate. Vascular infections usually cause wilting and discoloration as a result of the physical blockage of infected xylem vessels. True vascular wilt pathogens colonise the vascular tissue exclusively, although other pathogens can cause the same symptoms if they infect the vascular system as well as other tissues. There are a few pathogens that manage to achieve systemic infection of their host. For example, many viruses can spread to most parts of the plant, although not necessarily all tissues.

Some downy mildews can also systemically infect their host by invading the vascular tissue and growing throughout the host, causing deformation, rather than necrosis. Finally, there are some pathogens that complete their entire life cycle within the cells of their host, and may spread from cell to cell during cytokinesis. These are endobiotic infections.

DISEASE PHYSIOLOGY

While necrotrophs have little effect on plant physiology, since they kill host cells before colonising them, biotrophic pathogens become incorporated into and subtly modify various aspects of host physiology, such as respiration, photosynthesis, translocation, transpiration and growth and development. The respiration rate of plants invariably increases following infection by fungi, bacteria or viruses.

The higher rate of glucose catabolism causes a measurable increase in the temperature of infected leaves. An early step in the plant's response to infection is an oxidative burst, which is manifested as a rapid increase in oxygen consumption, and the release of reactive oxygen species, such as hydrogen peroxide (H_2O_2) and the superoxide anion (O_2^-). The oxidative burst is involved in a range of disease resistance and wound repair mechanisms.

In resistant plants, the increase in respiration and glucose catabolism is used to produce defence-related metabolites via the pentose phosphate pathway. In susceptible plants, the extra energy produced is used by the growing pathogen.

Pathogens also affect photosynthesis, both directly and indirectly. Pathogens that cause defoliation rob the plant of photosynthetic tissue, while necrotrophs decrease the photosynthetic rate by damaging chloroplasts and killing cells. Biotrophs affect photosynthesis in varying degrees, depending on the severity of the infection. A biotrophic infection site becomes a strong metabolic sink, changing the pattern of nutrient translocation within the plant, and causing net influx of nutrients into infected leaves to satisfy the demands of the pathogen.

The depletion, diversion and retention of photosynthetic products by the pathogen stunts plant growth, and further reduced the plant's photosynthetic efficiency. In addition, pathogens affect water relations in the plants they infect. Biotrophs have little effect on transpiration rate until sporulation ruptures the cuticle, at which point the plant wilts rapidly. Pathogens that infect the roots directly affect the plant's ability to absorb water by killing the root system, thus producing secondary symptoms such as wilting and defoliation.

Pathogens of the vascular system similarly affect water movement by blocking xylem vessels. Growth and development in general are affected by pathogen infection, as a result of the changes in source-sink patterns in the plant. Many pathogens disturb the hormone balance in plants by either releasing plant hormones themselves, or by triggering an increase or a decrease in synthesis or degradation of hormones in the plant. This can cause a variety of symptoms, such as the formation of adventitious roots, gall development, and epinasty (the down-turning of petioles).

INFECTION PROCESS

Plant pathogens fall into two broad categories: necrotrophs (those that kill plant cells before parasitising them), and biotrophs (those that obtain nutrients from living cells). Failure of pathogens to invade suitable host cells (dead in the case of necrotrophs; alive for biotrophs) will prevent them from infecting the host and the plant will be *resistant*. Additionally, the establishment of a parasitic relationship is dependent on the responses of the plant under attack.

Resistant hosts prevent or slow the development and

reproduction of the majority of pathogen propagules that they come into contact with. Resistance can be expressed at many stages in the infection process, from inhibition of propagule germination and penetration, to the restriction of colony development after the pathogen has become established. The defence barriers erected by plants are a co-ordinated system of molecular, cellular and tissue-based responses to pathogen attack.

DISEASE DEVELOPMENT

The amount of disease that develops in a plant community is dependent on properties of the host, the pathogen and the environment. The environment can affect both the susceptibility of the host (e.g. by creating stress in the plant) and the activity of the pathogen (e.g. providing moisture for spore germination). The pathogen and the host can affect each other's performance. The plant can also change its environment, by creating a microclimate around it.

Table: Factors that Affect Disease Development.

Pathogen	Host	Environment
• Presence of pathogen • Pathogenicity • Adaptability • Dispersal efficiency • Survival efficiency • Reproductive fitness	• Susceptibility • Growth stage & form • Population density & structure • General health	• Temperature • Rainfall / Dew • Leaf wetness period • Soil properties • Wind • Fire history • Air pollution • Herbicide damage

THE PATHOGEN

The presence or absence of a pathogen is the main factor that determines whether disease occurs. Introduction of a pathogen to an area from which it has previously been absent can cause major outbreaks of disease in plant communities.

The amount of disease that develops is often determined by the pathogenicity of the main pathogen.

The term pathogenicity relates to both the virulence (infection ability) and the aggressiveness (the vigour of the infection) of the pathogen. Pathogenicity is dependent on the pathogen's reproductive, dispersal and survival fitness.

The adaptability of the pathogen is also important in determining its ability to infect resistant hosts or to survive changed environmental conditions. Adaptability is determined by the pathogen's genetic flexibility and reproductive efficiency. The spread of a disease and the formation of epidemics is reliant on the pathogen's ability to disperse rapidly over long distances.

The spores of cereal rusts, for instance, can be blown over vast distances in a few days, while soil-borne pathogens have little scope for extensive spread. For a pathogen to cause disease in successive seasons, it must be able to survive the intervening time. Some pathogens form spores or sclerotia that can survive in the soil for years, while others colonise alternative plant species until the season of their primary host comes around again. Beyond the mere presence of a pathogen, the number of infective propagules available to infect plants is a crucial factor in determining the amount of disease that develops.

Generally, as the number of propagules increases, the level of disease increases, levelling off when the amount of disease reaches very high levels and there are few uninfected plants available. The survival of propagules, and therefore the number of propagules available to cause disease, is heavily influenced by environmental factors.

THE HOST

The development of disease in a plant community relies on the presence of individual hosts that are susceptible to that particular pathogen. If the majority of the population is susceptible to the pathotypes of a pathogen in the vicinity, an epidemic can occur.The best way of controlling disease is by planting species or cultivars that are not susceptible to pathogens of that area. The occurrence of disease can also be influenced by the host plant's growth stage and form. Some diseases are common in seedlings, while others are typical of mature plants. The growth stage of the population can also affect the microclimate around the plants; for example, the humidity and sunlight levels under the canopy. The population structure and density will also affect the development of disease in a plant community.

The density of the main host species and the proportion of other plants that are not hosts within the community will determine the rate and extent of epidemic development. Crop plants tend to be densely planted, with no other species in amongst them, making them more susceptible to rapid spread of disease. Extensive, dense plantations can host spectacular epidemics, particularly if a new pathogen is introduced to the area. In addition, the general health of the host plant before infection is important in determining the success of a disease. Necrotrophs do well on poorly growing plants, while biotrophs thrive on a healthy host plant.

THE ENVIRONMENT

The presence of a pathogen against a particular plant will generally not cause serious disease unless the environmental conditions are favourable. This includes the aerial environment and the soil (*edaphic*) environment.

Human attempts at controlling disease usually involve manipulating the environment in some way. For example, breeding wheat cultivars to tolerate dry conditions allows Australian farmers to plant the crop in areas that are not favourable for pathogens such as powdery mildew and leaf rust. Properties of the aerial environment that influence disease development include moisture levels, temperature and pollution.

Moisture is particularly important to pathogenic bacteria and fungi. Rain splash plays an important role in the dispersal of some fungi and nearly all bacteria, and a period of leaf wetness is necessary for the germination of most airborne spores. By using water for dispersal, propagules are dispersed at a time when they are likely to be able to germinate as well. Because the process of germination and infection takes time, the duration of leaf wetness also influences the success of the infection. The duration necessary for infection varies with temperature. Usually, a longer period of leaf wetness is needed to establish an infection in cooler temperatures, as germination and infection are generally accelerated in warmer conditions.

Temperature also affects the incubation, or latent, period (the time between infection and the appearance of disease symptoms), the generation time (the time between infection and sporulation), and the infectious period (the time during which the pathogen keeps producing propagules). The disease cycle speeds up at higher temperatures, resulting in faster development of epidemics.

The period of leaf wetness, combined with temperature information can be used to predict outbreaks of some diseases (infection periods) and be used to time preventative treatments, such as spraying. A recently recognised aspect of the aerial environment that can influence disease in plants is air pollution. A high concentration of pollutants can affect disease development and, in extreme cases, damage the plants directly by causing acid rain.

The edaphic (soil) environment affects soil-borne diseases, largely by determining the amount of moisture available to pathogens for germination, survival and motility. Germination and infection success also rely on the temperature of the soil.

The fertility and organic matter content of the soil can affect the development of disease. Plant defences are weakened by nutrient deficiency, although some pathogens, such as rusts and powdery mildews, thrive on well-nourished plants. Other diseases thrive in soils that are specifically low in organic matter.

INTERACTION BETWEEN FACTORS

The pathogen, the host and the environment interact, usually in ways that are difficult to quantify and predict. Control measures can include sowing of a crop species early, to avoid exposing seedlings to a disease during the time of year that provides the best environmental conditions for the pathogen.

EPIDEMIOLOGY

Prediction of disease outbreaks enables the effective use of control measures, such as chemical or biological treatments, the prediction of crop yields and of the market potential for that crop. Disease forecasting involves the use of weather data and biological data to predict disease incidence. Usually, disease forecasting is only performed on economically important diseases, and as a method of cost reduction.

If controlling a particular disease involves an expensive or time-consuming treatment, being able to predict outbreaks of the disease allows the treatment to be timed correctly, increasing its effectiveness, and reducing the cost compared to repeated treatments. Because environmental conditions vary from season to season, disease forecasting is necessary to predict the chance of disease in a certain set of conditions.

Disease can be forecast using computer modelling and empirical correlations relating to weather conditions, levels of inoculum, test plots and site factors and the predictions can then be communicated to growers. Computer modelling of plant diseases uses systems analysis to accumulate all the factors that affect the development of a certain disease into a computer-based model, and make predictions of disease under different environmental conditions.

A disease needs to be well understood in order to formulate an accurate model, and models based on diseases that we know little about are generally not very accurate. The more straightforward approach of developing empirical correlations between particular weather factors and disease has had considerable success. This does not attempt a complete modelling of all factors involved in a disease, but only those most important in affecting the disease. The accuracy of the model can then be measured statistically by comparing its predictions to what actually happens.

Monitoring the weather is the most important

consideration in disease forecasting, because of the overriding effect that weather has on disease development. While broad scale weather data has been used for disease forecasting, it is well known that the microclimate within the crop has a more direct impact on disease. Devices have been developed to monitor microclimate factors such as duration of leaf wetness and temperature, and with time, they will be affordable and accurate enough for widespread use on individual farms.

Synoptic weather forecasting charts can be used to predict 'critical periods' - the occurrence of conditions favourable for disease development - so that farmers can spray their crop before it happens. There are now several self-calculating disease forecasting monitors available commercially that use environmental data and past season data to predict outbreaks of particular diseases.

Some disease forecasting methods are based solely on monitoring inoculum levels, often as indicated by the amount of disease already present. This method can be successful when disease is developing steadily under relatively uniform or predictable weather conditions, but not for diseases that can spread explosively in favourable conditions. Monitoring the amount of disease present can indicate whether the amount of disease is likely to exceed a certain threshold, at which point control measures become economical.

There are numerous methods of directly monitoring the concentration of spores in the air as an indication of the chance of disease. Trapping vectors of diseases can also be useful in predicting the occurrence of viral diseases. In addition, estimation of populations of soil-borne pathogens by examining soil samples is necessary for predicting the outbreak of the diseases they cause. Monitoring systems can be combined with data specific to the site and the crop, such as soil type, topography and irrigation levels, in order to increase the accuracy of predictions.

Test plots (or trap plots) of susceptible cultivars can be planted throughout a cropping area to give early warning of the arrival of inoculum or disease vectors. Alternatively, inoculation of the test plot with the pathogen can give an indication of favourable environmental conditions for disease development. Test plots are also useful for monitoring the occurrence of minor diseases on new cultivars.

The formation of a prediction is useful only if it can be communicated to the growers who will be affected by it. General warnings for areas are broadcast over the radio or internet. Predictions based on monitoring by individual farmers or groups of farmers in an area remove the need for widespread communication systems. Computerised decision support systems based on local monitoring can educate and empower farmers when making decisions about their crops.

DISEASE ASSESSMENT

Critical information in the assessment of disease is the amount of disease that is present. This can be measured as the proportion of a plant community that is diseased (disease incidence) or as the proportion of plant area that is affected (disease severity). Often, disease has to exceed a certain threshold before it reduces the yield of a crop, but it is usually difficult to accurately estimate the yield reduction caused by a specific disease.

For example, many diseases occur on senescing tissue that would not have contributed to the yield anyway. Easier diseases to assess are those that kill whole trees in orchards or plantations, and those that destroy the actual harvested product, such as fruit or grain.

Disease and crop loss assessments are necessary for evaluating the economic impact of a disease and the benefit of particular control strategies. There is no point in implementing a control measure if it will cost more that the increased crop yield will return. The growth of the crop, its yield potential, the development of the disease and its impact on yield all have to be measured to predict the impact on yield of particular levels of disease. This information can be combined with predictions of likely disease levels to determine whether preventative treatments are worthwhile.

In the past, most crop loss assessment has been qualitative, producing vague, inaccurate and sometimes misleading data.

One major problem with this is the complex nature of disease development. Rarely can disease be attributed to just one factor. Assessment of the effect of disease on crop yield normally involves five steps:

- Developing a descriptive growth stage key for the particular crop species
- Developing methods to assess the incidence and severity of disease
- Developing statistically sound methods of sampling crop populations for assessment of the amount of disease
- Estimating the negative impact of particular levels of the disease on crop yield and quality, and
- Evaluating the economic benefit from various methods available for reducing the amount of disease.

ASSESSMENT OF CROP GROWTH AND DEVELOPMENT

The first step to quantify the effect of disease is to develop a key that describes the growth and development of healthy plants during the growing season. It should describe development, either from sowing to harvest, in the case of annual plants, or from season to season in perennial plants.

Details drawings or photographs, showing characteristics of the various stages of development, including leaf formation, flowering, fruiting and senescence are needed. Standardised growth keys have been developed for a number of crop plants, enabling comparison between different countries and different conditions.

ASSESSMENT OF DISEASE INCIDENCE AND SEVERITY

Disease assessment methods need to easily provide objective measurements for a specific growth stage of a crop so that data from different sources is comparable, and provide an adequate sample of the crop for assessment. Whether it is disease incidence, or disease severity, or both, that are measured, depends on the nature of the disease. An "all or nothing" disease, for example, that inevitably kills any plant it infects, could be measured just by counting the number of plants that infected (disease incidence).

However, in the case of a disease that causes varying degrees of damage to plants throughout the crop, a more complex measurement is needed, that assesses disease severity.The disease incidence for biotrophic pathogens can be measured by counting the number of plants, leaves, flowers etc that are infected, but the disease severity is assessed by estimating the proportion of total photosynthetic area that is diseased. While this is generally less precise and less controllable than counting individual plants, it is usually a better predictor of crop loss. Because judging the proportion of diseased leaf by eye is unreliable, disease assessment keys, showing different disease severities as blackened areas, have been devised for various crops.

To produce a disease assessment key, the development of disease over the whole disease cycle and at different stages of plant growth must be studied to make prototype standard diagrams and/or descriptions. The accuracy of the key then needs to be tested, by assessing disease severity in the field using the key, and then assessing the same samples using accurate measurement techniques in the laboratory.

There are also computer-programs designed to train observers in disease severity assessment, by presenting images of diseased leaves, which the observer assesses, and comparing their result with the known level of disease in the diagram. This aims to reduce variation in results caused by different observers.

These disease assessment methods will only be accurate if performed on a representative sample of the crop. Samples of crop units (plants, leaves, fruit etc) can be taken randomly from a crop, or standard quadrats can be placed in the crop and all plants within the quadrat assessed. In a test plot, a part of the plot is usually assessed for disease. Taking samples only from the edge of a plantation will not necessarily be representative of the bulk of the crop.

To determine how many samples need to be taken, it is possible to sample a number of times with progressively more samples, and find the point at which the standard error is low. A disease that is uniformly spread throughout a crop will require fewer samples for an accurate assessment than a disease that has a patchy distribution throughout the crop.

ASSESSMENT OF CROP LOSSES

Once the amount of disease has been determined, the next step is to assess, either experimentally or statistically, the effect of different levels of disease on the crop yield. Experimental assessment involves setting up test crops in which the level of disease is controlled. Monitoring crops with different levels of disease allows the comparison of epidemic progress and crop yield under different disease conditions.

A relationship between disease parameters and yield can then be formulated, allowing prediction of crop loss for a certain level of disease at a particular point in the crop's growth. This method is dependent upon the assumption that the treatments used to keep disease at a certain level have no effect on crop yield themselves.

This might not always be the case. Similarly, assessments using susceptible and resistant cultivars rely on both cultivars having a similar yield to start with. Again, this might not be true, and must be determined fist in a disease-free environment. All aspects of experimental design need to be carefully considered in order to gain an accurate assessment of the effect of disease on crop yield.

For example, harvesting methods under experimental conditions are often more efficient than under field conditions, giving the appearance of a higher crop yield. From crop loss assessment studies, models can be devised. Crop loss moels are usually based on one of three types of disease assessment: disease at a critical point in development, disease at multiple points in development, and disease throughout crop development.

The statistical approach to assessing disease involves statistical analysis of crop yields under different levels of disease that occur naturally in the field. The levels of disease and the crop yields are monitored, but not manipulated, and then yields from different seasons or areas with different levels of disease can be compared to determine the effect of disease on crop yield. The advantage of this method is the use of outcomes of real cropping situations in the field, not experimentally manipulated crops.

The disadvantage of this approach is that in the field, the conditions are not controlled, and there are many factors that could affect yield besides disease. An alternative statistical approach to crop loss assessment is the use of questionnaires, filled in by the farmers, that allow disease and pest incidence and severity to be related to the yields they produce.

DISEASE MANAGEMENT

All plant diseases result from a three-way interaction between the host, the pathogen and the environment. An epidemic develops if all three of these factors are favourable to disease development. Therefore, disease can be controlled by manipulating one or more of these factors so that conditions are unsuitable for replication, survival or infection by the pathogen. The pathogen can be manipulated by excluding or reducing inoculum available to infect plants. If a particular pathogen is not present in an area then attempts can be made to prevent its introduction. Quarantine regulations govern the introduction of plant material from high-risk areas. Seed certification schemes also reduce the risk of introducing a new pathogen, and propagating material can be treated to kill any inoculum that is present, before it is introduced to a new area.

If a pathogen is already present in an area, efforts to eliminate or reduce the amount of inoculum present are the only option. Elimination of an existing pathogen is usually not successful, but there are a variety of methods, such as crop rotation, chemical and physical treatment and trapping, that can effectively reduce the amount of inoculum available. The nature of the pathogen can have a significant impact on the management of its disease.

The host can be manipulated by increasing its resistance to disease. This can involve breeding or selecting plants for genetic resistance to a specific disease, or by applying chemicals, such as fungicides, that prevent or halt infections in the plant.The environment can be modified so that it is suitable for plant growth but not for disease development. For example, improving soil drainage, changing the time of sowing, reducing the density of plants in a crop or changing irrigation practices can all produce conditions unfavourable to particular diseases, while maintaining conditions suitable for the growth of the plant. Any decrease in the amount of initial inoculum, the rate of disease increase of the time available for development, will cause a decrease in the amount of disease. Therefore, management strategies look at reducing one or more of these factors in order to reduce the amount of disease. The initial amount of inoculum can be controlled by quarantine practises and preventative treatments.

The time available for disease development can be reduced in some kinds of crops by harvesting before the disease has developed to epidemic proportions. The rate of disease increase can be reduced by mixing plants that are not all susceptible to the same strain of pathogen, by applying preventative chemical treatments, and by changing the environments the host and pathogen are exposed to, for example, by changing planting dates, crop density or irrigation practices. Integrated pest management uses all suitable techniques that complement each other with the aim of keeping pest populations below the threshold at which economic damage occurs. This approach also aims to avoid the problem of pests developing resistance to widely used insecticides.

In integrated pest management, resistant crop varieties might be used in conjunction with chemical treatment, crop rotation and manipulation of the environment. Multiple approaches to disease management are particularly useful when the disease cycle is not fully understood.

QUARANTINE

EXCLUSION, ERADICATION AND ELIMINATION OF DISEASE

Most pathogens of economic and biological significance

in Australia have been introduced. The prevention of pathogens entering new regions is known as exclusion, and is achieved by quarantine or by treating propagating material (elimination) before its introduction. If an outbreak of a disease occurs in a new area, efforts are made to eradicate the pathogen from that area.

EXCLUSION

The serious consequences of past outbreaks of disease, both in humans and in commercially important plant species have ensured that quarantine is taken very seriously in Australia, where it is administered by the Australian Quarantine and Inspection Service (AQIS).

While total bans have been put in place in the past, they generally lead to increased illegal smuggling of the banned item. Instead, quarantine aims to prevent entry of dangerous pathogens, but not by completely blocking movement of biological material. The quarantine service determines whether the threat posed by a proposed import is an acceptable risk; that is, the risk is small enough to be manageable.

A crucial process in quarantine regulation and the co-operation of countries in the International Plant Prevention Convention, is risk identification, assessment and management. Pest risk analysis addresses the injury or potential injury that a plant, animal or pathogenic agent can cause in an area. Risk identification determines whether the organism in question qualifies as a quarantinable pest.

Risk assessment establishes the probability of the pest being introduced to the country, for example by wind, or via a vector species, and the chance that the pest will become established once introduced. Risk management aims to reduce the risk of introduction and establishment of pest species, for example by fumigating biological material when it enters the country.

THE SPECIAL CASE OF AUSTRALIA

The fact that Australia is an island, much of which is inhospitable to many organisms, acts as a natural defence against the introduction of exotic pests and diseases. However, it also means that Australia's native plant and animal species have evolved without exposure to many pathogens, making them particularly vulnerable to introduced pathogens. Northern Australia is at particular risk because of its proximity to Papua New Guinea and Indonesia, and the trading routes that exist in the area.

Quarantine practises are also in place within Australia to prevent the spread of pathogens from one area of the country to another. For example, particular fruits cannot be brought into major fruit growing regions that are free of pathogens that affect their main crop.

The introduction of a new pathogen to a fruit-growing area has the potential to devastate the entire industry in that area, and so quarantine regulations are strictly enforced. If a new disease does occur, procedures need to be carried out to prevent its spread. Containment and removal of infected plants is essential, and where possible, plants are also treated to destroy the pathogen.

SANITATION

Sanitation involves all procedures that prevent the spread of disease to new plants, plant products, and new areas, or to reduce the amount of inoculum in an already affected area. It includes thorough washing or chemical treatment of machinery, tools, potting equipment, shoes and hands that will come in contact with multiple plants, washing produce and its containers and storage areas, protecting soil used for propagation from contamination.

Removing and disposing of infected leaves and other plant material is also crucial for reducing the amount of inoculum available to cause new infections.

DISEASE-FREE PROPAGATING MATERIAL

Propagating material includes seeds, cuttings, root-stocks, buds, tubers, rhizomes, bulbs and corms. All of these can carry pathogens, so it is essential that only disease-free propagating material is introduced to new areas. This can be achieved by producing the propagating material in disease-free areas, inspecting the material carefully before it enters a disease-free area and treatment of or destruction of any diseased material that is found.

The material can be inspected at any number of stages between production and planting, and again after planting to check that no disease has developed. Potentially diseased material can be successfully treated with heat for some pathogens, as long as the host heat tolerance level is higher than that of the pathogen. This is a particularly useful method for infections that are deep in the tissue and can't be reached with chemical treatment, such as viruses and bacteria.

Alternatively, seeds can be treated with chemicals that kill pathogens on their surfaces. There are also certification schemes that ensure that seed and vegetative propagating material are free from particular diseases.

ERADICATION

Eradication refers to the process of removing all infected plant material in an area, usually involving the destruction of large numbers of plants in infested areas. Some diseases have been effectively eradicated after outbreaks in the past, but it is not always successful.

The time between introduction and identification of a disease in a new area can allow inoculum to accumulate to a point where eradication or even containment is impossible. This makes effective quarantine practises especially important. As well as destroying the infected plants and leaf litter, soil and water may need to be sterilised to prevent the pathogen spreading.

Steam can be used to pasteurise soil and potting media; chemicals, such as methyl bromide can be used to fumigate them, or solarisation, using heat from the sun can also raise the temperature of soil or potting mixture to levels high enough to kill pests and pathogens.

Irrigation water can be treated to remove the propagules of pathogens by filtration, pasteurisation with heat, irradiation with UV light, chlorination, bromination (the use of bromine), or ozonation (the use the strong oxidising agent ozone). Of these, chlorination is generally the cheapest and most reliable.

CHEMICAL CONTROL

Chemical disease control employs the use of chemicals that are either generally toxic and used as disinfectants or fumigants or chemicals that target specific kinds of pathogens, as in the case of fungicides, bactericides (or antibiotics) and nematicides. No substances have yet been found that effectively control plant viruses.

Ideally, a chemical control agent should be effective at concentrations that will not harm the plant, have low risk to humans and animals, and have minimal effect on the normal microflora on the plants and in the soil. Also, there should be little chance of the pathogen quickly developing resistance to it, and it should be suitable for long periods of storage in ambient conditions. These chemical agents can be sold as dusts, concentrated solutions, wettable powders, granules or emulsions.

FUNGICIDES

Fungicides are chemicals used in the control of fungal diseases. They are often classified as either protectant or systemic. Protectant fungicides are usually effective against a broad range of fungi and protect the plant against infection on the surfaces of the plant to which they are applied. Often, they require multiple applications during the growing season to maintain coverage as new growth emerges and weathering removes past coverage.

Systemic fungicides can be absorbed by the plant without harming it, and transported to other tissues where they are toxic to fungi. These compounds can control and eradicate established infections, but they are also vulnerable to fungi developing resistance, as they generally only target one step in a biosynthetic pathway to kill the fungus.

To minimise the development of resistance by chemical overuse, fungicides are classified into groups based on their chemical activity. By alternating between the different classes of fungicides, the fungal population has less opportunity to build up resistance to one chemical.

ANTIBIOTICS

Relatively few antibiotics are routinely used to control plant diseases. Antibiotics are chemical produced by micro-organisms, which destroy or injure living organisms, in particular, bacteria. Streptomycin is effective against a few fruit pathogens, such as blights and cankers, and cyclohexamine can be used to control some fungal pathogens of crops, particularly powdery mildews and rusts. Bacteria, as well as fungi, have the ability to develop resistance to antibiotics, which is a major disadvantage of using these compounds, and one of the reasons that they are not widely used.

NEMATICIDES

The use of nematicides is confined largely to high-return horticultural crops, because they are expensive. Additionally, they are all highly toxic, and alternative measures for controlling nematodes are being investigated.

Fig. Nematode Damaged Wheat Field

It estimated that there are nearly one million species of nematodes, making them the second most common animals on earth, after insects. They are found virtually everywhere and are responsible for an estimated eighty billion dollars worth of crop damage and loss throughout the world each year.

Nearly all crops are damaged by at least one nematode species, however low-level damage often remains undetected despite reduced crop quality and quantity. Some common nematode pests include root knot nematode, cyst nematode, citrus nematode, lesion nematode and stubby root nematode.

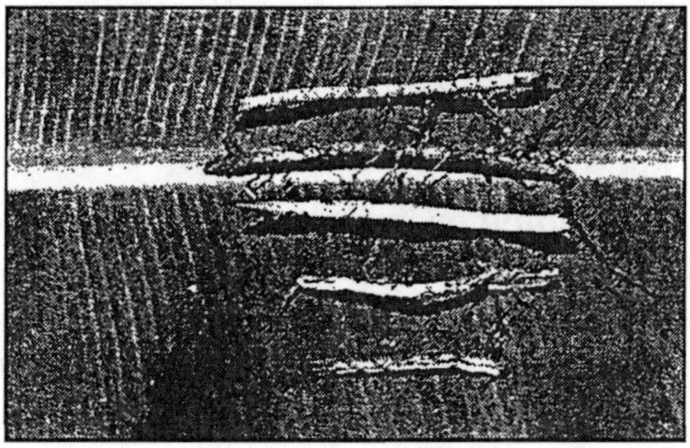

Fig. Nematode Damaged Roots

Chemical nematicides are highly toxic and have the potential to be seriously environmentally damaging which has led to research in alternative methods for nematode control. One of the most promising solutions is the use of nematode-trapping fungi, commercially known as myco-nematicides.

These fungi are able to form organs on their hyphae which trap the nematode. There are two types of traps, non-adhesive and adhesive. Non-adhesive traps may include constricting rings which spring closed and capture the nematode, while an adhesive trap may include a sticky network of filaments in which the nematode is caught and then digested by the fungi.

CHEMICAL DISEASE MANAGEMENT

Pathogen resistance to chemical agents has been a problem since the 1970s. Throughout the 1980s and 1990s, public concern has grown about the use of agricultural chemicals in general.

The main concerns are the risk of poisoning humans or animals, contamination of livestock products, harm to beneficial insects, and the contamination of food products, waterways and soil. When properly used, fungicides should not cause problems in any of these areas, but they do have the potential to alter the balance of diseases on certain crops.

The main risk to humans is during the preparation or application of these chemicals, when they can be inhaled, ingested or absorbed through the skin, and upon consumption of plants or their products. To overcome these risks, safety precautions must be followed when working with the chemicals, and various regulations exist with regard to consumable plant products.

For example, some chemicals may not be applied within a certain time before harvesting (the with-holding period), thus ensuring that chemical residues have fallen to an acceptable level by the time they reach the market. Maximum residue limits have been established to allow monitoring of the chemical residue on plant products.

Environmental concerns focus mainly on protectant fungicides. For example, copper and sulphur sprays have the potential to affect a broad range of organisms if they are washed off the leaves and accumulate in the soil or are washed into the waterways.

In response to public concern about these issues, many countries have initiated programs to reduced pesticide use. They can involve education of farmers on best practices in pesticide application, the use of lower doses of chemicals, better understanding of the threshold level of infection that causes economic loss and developing disease forecasting systems to allow optimal use of chemicals.

BIOLOGICAL CONTROL

Biological control of disease employs natural enemies of pests or pathogens to eradicate or control their population. This can involve the introduction of exotic species, or it can be a matter of harnessing whatever form of biological control exists naturally in the ecosystem in question.

The induction of plant resistance using non-pathogenic or incompatible micro-organisms is also a form of biological control.

Some diseases that can be successfully controlled using biological agents are pathogens of pruning wounds and other cut surfaces, crown gall, diseases of leaves and flowers, such as powdery mildew, diseases of fruits and vegetables, such as *Botrytis*, and fungal pathogens in the soil (disease suppressive soils).

NATURAL CONTROL

All biological communities are complex ecosystems in which the abundance of any organism is dependent on its food supply, its environment and other organisms. If a pathogen is kept in check by the microbial community around it, then biological control has been achieved.

Biological control appears to take place on the plant surface by the activity of epiphytic microflora. This is then an important consideration when applying chemicals to plants, since there is a risk of killing natural antagonists of pathogens other than the one being treated.

HOW BIOLOGICAL CONTROL AGENTS WORK

The most common mechanisms for microbial antagonism of plant pathogens are parasitism, predation, competition, induced resistance and the production of antimicrobial substances. Often, several mechanisms act together.

Competition exists between organisms that require the same resource for growth and survival. Use of the resource by one organism reduces its availability for the other organism. Competition for space or nutrients usually takes place between closely related species. Therefore, it can be effective to treat plants or seeds with a non-pathogenic strain of a related species that can out-compete the pathogenic organism. In some cases, the treating species need not be closely related to the pathogen, as long as it uses the same resources. For example, bacteria and yeasts can reduce fungal spore germination by competing with the spores for nutrients on the surface of leaves.

While micro-organisms can produce secondary metabolites that have anti-microbial properties when grown in culture, these chemicals are rarely detected in natural environments. Therefore, antibiotics would need to be produced in culture and then applied. However, antibiotics are easily lost to the atmosphere and are commonly broken down by organisms that are insensitive to them, and so they are not ideal biological agents against plant pathogens.

Parasitism of one fungus by another (*hyperparasitism* or *mycoparasitism*) is well documented and is affected by environmental factors, including nutrient availability. Formulations of some parasitic species of fungi are available commercially for the control of fungal plant pathogens in the soil and on the plant surface.

The hyphae of parasitic fungi penetrate their victim, sometimes with the aid of wall-degrading enzymes. Bacteria on the plant surface and in the soil are also known to parasitise plant pathogens, such as other bacteria and fungal spores. Predation of plant pathogens by invertebrates can also contribute to general biological control. Bacterial feeding nematodes consume large numbers of bacteria in the soil and some amoebae are known to attack yeasts, small spores and fungal hyphae, although these organisms are generally non-specific predators and their relative importance in biological control is not well understood.

Induced resistance and cross-protection are two mechanisms of plant 'immunity' against a pathogen. In the case of cross-protection, an organism present on the plant can protect it from a pathogen that comes into contact with the plant later. For example, symptomless strains of tobacco mosaic virus can protect tomatoes from virulent strains of the same virus, rather like immunisation in animals.

Induced resistance is a form of cross-protection, where the plant is inoculated with inactive pathogens, low doses of pathogens, pathogen-derived chemicals or with non-pathogen species to stimulate an immune response. This prepares the plant for an attack by pathogens, and its defence mechanisms are already activated when infection occurs. It provides protection against a wide rage of pathogens across many plant species.

COMMERCIAL USE OF BIOLOGICAL CONTROL

Commercial application and grower acceptance of biological control has been slow to develop, mainly due to the variation in efficacy under the range of environmental conditions likely to occur in the field. This problem can only be overcome by better understanding the environmental parameters that limit biological control.

In addition to this problem, there has been relatively little investment in the development of commercially viable products for biological control, partly due to the cost of developing, testing efficacy and risk, registering and marketing such a product.

The most successful product would be one that can be applied using existing machinery or methods. Biological control agents are therefore generally formulated as wettable powders, dusts, granules and aqueous or oil-based liquid products, with various additives to attain all the desirable attributes.

THE FUTURE OF PATHOGEN CONTROL

While it is unlikely that biological control will completely replace chemical pesticides in the foreseeable future, we can expect that there will be some decline in the use of chemicals, particularly in developed countries.

Thus far, most approaches have involved the single antagonist concept, although a biological systems approach, where disease is suppressed from several angles, might provide a better alternative. Similarly, the use of biological control agents could be used as one component of an integrated management programme to achieve the best possible results.

Cultural Disease Management Practices

Cultural disease management practices are the measures undertaken by humans to prevent and control disease by manipulating plants. In the case of low-return crops, these might be the only forms of disease management that are economically viable.

Cultural management can include reducing the amount of initial inoculum, reducing the rate of spread of an established disease, or planting a crop at a site that is not favourable to pathogens because of its altitude, temperature, or water availability.

REDUCING THE LEVELS OF INOCULUM

Practices that reduce the initial levels of inoculum include selecting appropriate planting materials, destruction of crop residues, elimination of living plants that carry pathogens, and crop rotation. The selection of appropriate planting material can involve planting resistant cultivars, planting a number of mixed cultivars, using certified seed and ensuring that disease is not spread on vegetative propagating material or on equipment.

Many agricultural systems are characterised by dense populations with genetic homogeneity (monocultures). Once a disease becomes established in such a plant community, it can rapidly spread to epidemic proportions. Hence the value of planting mixed cultivars, which incorporate a range of resistances.

The destruction of crop residues, which can harbour many pathogens, by burying, burning or removal is an important practice performed between cropping seasons. How this is done depends on the type of crop, the type of pathogen and the size of the crop. Burying crop residues can destroy some pathogens, if ploughed in deeply enough, but some pathogens can survive, and even benefit from this process if it serves to spread them throughout the field.

Burning crop debris is a common practice, especially for cereal crops, and it is a successful method of destroying many pathogens. Its success depends on the intensity of the fire. There are disadvantages to burning also, such as loss of nutrients, smoke pollutions, increased soil erosion and contributing to the greenhouse effect. Because burning can completely destroy a source of inoculum, it is commonly used in eradication campaigns against newly introduced pathogens.

The elimination of living plants that carry pathogens can include both remnant or diseased crop plants and also wild plants or weeds that can act as alternative hosts between seasons. Some rusts, for example, cannot complete their life cycle in the absence of an alternate host, where they undergo sexual recombination. Removing alternate hosts can delay an outbreak, but often inoculum finds its way to crop plants via wind or another vector.

Crop rotation refers to the successive planting of different crops in the same area, sometimes with a fallow, or resting, period in between crops. The most successful practice rotates crops over periods that are longer than the survival period of the pathogen, thus reducing or eliminating inoculum in the interval before a susceptible crop is planted there again. It tends not t be very effective against pathogens that survive for a long time in the soil.

REDUCING THE RATE OF SPREAD OF DISEASE

The rate of disease spread can be reduced by controlling the spacing of plants, humidity, moisture levels, and amount of sunlight. These factors are more easily controlled in a greenhouse environment, but some can be manipulated in the field also. In general, wider spacing of plants reduces the speed with which disease moves between them, and reducing moisture levels can inhibit infection by pathogens.

The amount of time that the plants and soil are wet can be reduced by watering in the morning, spacing plants further apart, pruning or training to reduce canopy cover, and orienting rows of plants with relation to the prevailing wind. The use of shade cloths can help to control fungi that need UV light in order to sporulate.

Tillage practices have indirect effects on the spread of plant pathogens, although, some forms of inoculum can be spread extensively during tillage. Tillage can bury pathogens in the topsoil in deeper where they are less likely to cause disease. Preparation of seed beds can greatly alter the soil's texture, aeration, temperature, moisture levels and density. Tillage can also influence nutrient release in the soil and can generally benefit the crop.

Farming practices have moved away from regular tillage, reducing damage to roots and spread of pathogens caused by tilling machinery. However, minimum tillage can also encourage some pathogens, such as those that feed on crop residues left on the surface of the soil.

Sowing practices, such as changing time, depth and direction of sowing, and changing the density of the crop can protect plants from pathogens to which they are susceptible only at certain stages of their development. Changing the time of sowing can exploit weather conditions that are unfavourable to the pathogen, thus reducing crop losses.

This might require the use of a specific cultivar that is adapted to the selected growing period, but might also be susceptible to different pathogens. The depth of sowing can have a bearing on the chance of infection, as the seedling's pre-emergence stage, which is usually more susceptible to attack, is longer when seeds are planted deeper. However, deeper planting can stimulate germination.

Crop density and disease incidence are usually correlated, mainly due to the ease with which inoculum is transferred between plants when they are close together and their leaves and roots are able to touch. Also affecting plant susceptibility in densely planted crops is the microclimate created by their crowding.

Temperatures are more uniform, humidity is increased and leaves stay wet for longer, all of which favour the development of disease.Crop density can be manipulated by sowing, pruning, thinning, trellising, fertilisation, water management, staking and harvesting some plants or plant parts.

INTERCROPPING

The practice of planting more than one crop in alternating rows, or intercropping, can reduce disease by increasing the distance between plants of the same species, and creating a physical barrier between plants of the same species. Intercropping is more labour intensive the more crops there are, but it is usually beneficial.

How successful intercropping is depends partly on the combination of crop plants chosen, since some combinations can actually make disease worse by providing an alternate host or a stimulus that encourages germination of inoculum on the neighbouring species.

MULCHING AND SOIL AMENDMENTS

Mulches are used to conserve moisture and organic matter and reduce erosion in the soil. They are usually some kind of organic matter, such as straw, sawdust, manure or aquatic plants. Some manufactured products, such as plastics, asphalt paper and paper are also used. If crop residues are used for mulch they can provide a food source and attractive environment for pathogens and increase the incidence of a disease. Mulch or crop residues can also influence disease incidence by altering the immediate environment. Water retention, nutrient enrichment of the soil, decrease in soil temperature, weed inhibition and seedling protection are all effects of mulching that can influence disease development. Spread of soil-borne diseases that rely on splash-dispersal can be reduced, but other diseases, for which the altered environment is favourable, can be enhanced.

The addition of organic amendments to the soil can also reduce disease influence by increasing the activity of competing or predatory micro-organisms in the soil, although it can also increase the incidence or severity of diseases caused by pathogens that thrive on the amendment.

Flooding

Flooding can be used as a form of disease management, as in the example of growing rice in paddy fields. Its primary purpose is to reduce weeds, but it can also reduce the number of fungal propagules, insects and nematodes in the soil.

While flooding can aid the destruction of crop debris carrying inoculum, it can also carry propagules of some pathogens growing in the flood waters. Flooding has variable success in disease management, depending on the pathogens present.

Irrigation

Most forms of irrigation play a detrimental, rather than

beneficial role in plant disease management. For example, irrigation during dry seasons prevents the desiccation of propagules during drought, thereby increasing the level of inoculum. The irrigation water can also carry propagules and spread disease, unless it is treated before use.

Similarly, overhead watering can prolong leaf wetness, increasing the likelihood of germination and infection by fungal spores. It can also facilitate splash-dispersal of inoculum. However, irrigation does not always encourage disease, and in some cases it can be used to actively reduce the level of inoculum and retard disease development.

Alternate drying and wetting of soil encourages the activity of micro-organisms that destroy sclerotia, and brief overhead watering can wash spores away without causing long periods of leaf wetness. Trickle or drip irrigation is generally the least likely to encourage disease, since water is delivered directly to root zones and at a rate insufficient to disperse pathogens.

Roguing

The removal and destruction of diseased plants, or rogues, can prevent further spread of disease, but is labour intensive. For that reason, it is practicable only in small plantings, or where labour is cheap or if the crop is very valuable. Effective roguing requires the removal of diseased plants as soon as possible after symptoms are observed. In some cases, the dis ased plants as well as those immediately surrounding them, are destroyed in an attempt to halt spread of the disease.

Fertilisers and Crop Nutrition

A well-balanced supply of soil nutrients will result in healthy, vigorous plants, which should have a greater chance of withstanding attack by pathogens that unhealthy plants would. However, many pathogens also thrive under ideal growth conditions, particularly biotrophic pathogens, such as viruses. The major nutrients that influence plant and pathogen success are nitrogen, phosphorous, potassium and calcium.

The application of nitrogenous fertilisers delays crop

maturity by prolonging vegetative development. This can increase the risk of infection, since plants are more susceptible to disease at that stage of development. High nitrogen levels could also influence the production of metabolites by the host, with various outcomes for disease development. In some cases, infection is enhanced, in other cases it is inhibited. The addition of nitrogen to the soil can also alter the activity of micro-organisms and change the micro-environment in the crop due to increased canopy cover and crowding of foliage.

The addition of phosphorous fertilisers affects different crops and diseases in different ways, sometimes encouraging and sometimes inhibiting disease. The mechanisms through which phosphorous influences plant disease are not well understood. Potassium generally inhibits disease development, counteracts some of the disadvantages of nitrogen fertilisers, and promotes the healing of wounds in plants, all of which reduce disease.

However, potassium's effects can be variable, directly stimulating or inhibiting the penetration, multiplication, survival and establishment of a pathogen. Calcium is a necessary nutrient for the composition of plant cell walls. An adequate supply of calcium produces cell walls more resistant to penetration by facultative pathogens. High levels of calcium can also raise the pH of the soil, disadvantaging any pathogens that favour acidic soils. However, soils high in calcium can also promote the development of some diseases.

Essentially, the use of fertilisers has mixed effects on plant disease. Which fertiliser can be used without promoting disease will depend on the starting soil conditions and, importantly, the pathogens that are present.

Strip Farming

Strip farming is a similar practice to intercropping, where areas of one crop are separated from each other with strips of another crop, the rationale being that the two different crops are unlikely to share the same pathogens, thus reducing the rate of spread. The main disadvantage of strip farming is the fact that it is relatively labour intensive.

Trap and decoy crops

Trap crops of susceptible plants are grown on land known to contain pathogens. They become infected and are then destroyed before the pathogens' life cycles are complete, thus reducing the amount of inoculum in the area.

Decoy crops stimulate the hatching or germination of pathogen eggs or propagules, but the pathogens are unable to establish and infection of the decoy host and die, again reducing the amount of available inoculum. The ideal decoy crop is one that has economic value and can be used in a routine crop rotation.

Chapter 4

Pathogen Recognition

Plants use a vast array of signals originating from micro-organisms and the environment to recognise pathogens and elicit plant defence responses. Non-specific elicitors of biotic and abiotic origin induce host defences in a broad range of host species. Abiotic elicitors such as heavy metal ions or UV light can induce stress responses in exposed tissues, which may provide an additional barrier to invading pathogens or alternatively, increase the plant's susceptibility to infection.

Biotic elicitors include cell wall fragments released from fungi and bacteria, hydrolytic enzymes of plant or pathogen origin, some peptides, glycoproteins and polyunsaturated fatty acids. These elicitors induce defence responses in a range of host species. Often, non-specific elicitors act as a general indication that the cell has been damaged in some way (for example, the release of fragments of the host's own cell wall can elicit defence responses).

Specific elicitors enable defence against a very specific pathogen, and are conditioned by *avirulence genes in that pathogen*. Avirulence genes determine the pathogen's host range, but are only able to function in the presence of another set of genes, the 'hypersensitive response and pathogenicity' (HRP) gene cluster. Some Hrp gene products are involved in disguising the pathogen from host recognition, thus playing a role in both virulence and avirulence.

For a biotroph to form a successful infection, it must establish a basic compatibility with its host. The pathogen may also produce compatibility factors that delay, avoid or negate recognition by a normally resistant host plant. Virulent strains appear to be able to suppress the resistance mechanisms of the host, but are not able to halt resistance responses once they are activated. Incompatibility between a host and a pathogen results in the recognition of the pathogen and activation of defence mechanisms, while compatibility results in infection.

Specific elicitors are encoded by avirulence genes, and these peptides are believed to bind to receptor peptides, encoded by host resistance genes. Recognition of the avirulence gene products by the host triggers signal transduction pathways that cause a massive shift in gene transcription and plant cell metabolism, and local and systemic signals are released that prime the rest of the plant against further infection. The presence of non-specific elicitors, such as the release of host and pathogen wall fragments, during this process may amplify the defence response.

Host-parasite specific resistance is determined by the interaction of between products of pathogen avirulence genes, Specific elicitors and products of host resistance genes. The defence responses of plants can be very rapid. Host gene expression begins within minutes, or even seconds, of exposure to elicitors or pathogens. A diverse range of elicitors can induce a common set of responses in the host, suggesting that second messengers are involved in the signalling pathway between pathogen attack and host response.

RAPID ACTIVE DEFENSES

Almost every host-parasite interaction is unique in the details of the activation, localisation, timing and magnitude of the defence responses.

At the Membrane

The host membrane appears to be involved in the earliest stages of pathogen recognition and signal transduction. A change in membrane permeability after exposure to a pathogen causes fluxes in ions, such as K^+, H^+ and Ca^{2+}, and results in changes to gene activation and the triggering of the defence responses.

Also at the membrane, the 'oxidative burst', which involves the generation of reactive oxygen species, such as hydrogen peroxide, triggers signals that affect gene expression, cross-linking in the host cell wall and initiation of later defence responses. The reactive oxygen species at the site of infection are also produced in quantities capable of killing micro-organisms directly.

At the Cell Wall

Preparations for the reinforcement of the cell wall, which can improve host resistance, begin very quickly after a pathogen attempts to penetrate a host cell. This is characterised by an intensification of cytoplasmic streaming and the accumulation of host cytoplasm around the site of attempted penetration. The cytoplasmic aggregates are thought to contain cellular apparatus for the synthesis of cell wall fortifications.

If the host cell can repair and reinforce its cell walls quickly enough, it might reduce the penetration efficiency of the pathogen. Several types of reinforcement are produced by host cells. A papilla is a deposit of callose, silicon, lignin and proteins between the cell wall and cell membrane, directly below the point of attempted penetration, while lignitubers are lignified callose reinforcements that ensheath invading hyphal tips.

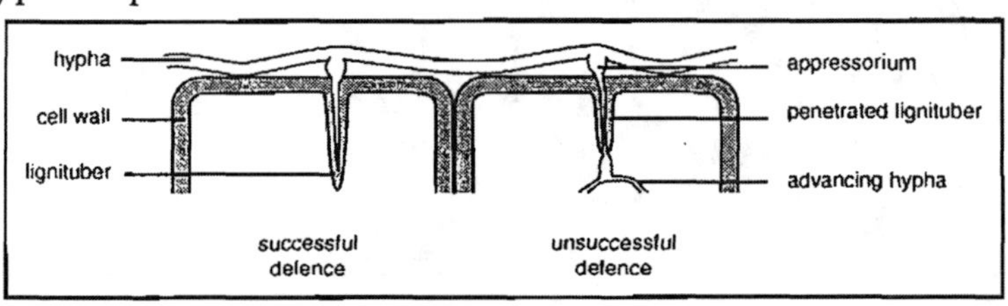

Fig. Lignitubers

This figure illustrates the difference between successful and unsuccessful defence against a fungal pathogen. In successful defence, the lignituber contains the fungal hypha, preventing further growth. In unsuccessful defence, the hypha is able to penetrate the lignituber and continue its growth into the plant.

Hydroxyproline-rich glycoproteins are structural cell wall proteins involved in secondary cell wall thickening. The expression of genes governing their production is activated ahead of invading hyphae, reinforcing walls. Cross-linking of hydroxyproline-rich glycoproteins caused by the release of hydrogen peroxide in the oxidative burst also reinforces cell wall compartments. Rapid deposition of lignin and suberin following infection also increases resistance to pathogens in many plants.

Lignin can also bind to hyphal tips and bacteria, physically restraining them and restricting the diffusion of their enzymes and toxins into, and the extraction of water and nutrients out of, the host cell. Cell wall reinforcements tend to be larger and more quickly formed in resistant hosts than in susceptible hosts, and inhibition of the production of callose or lignin synthesis by the pathogen enhances its penetration efficiency.

THE HYPERSENSITIVE RESPONSE

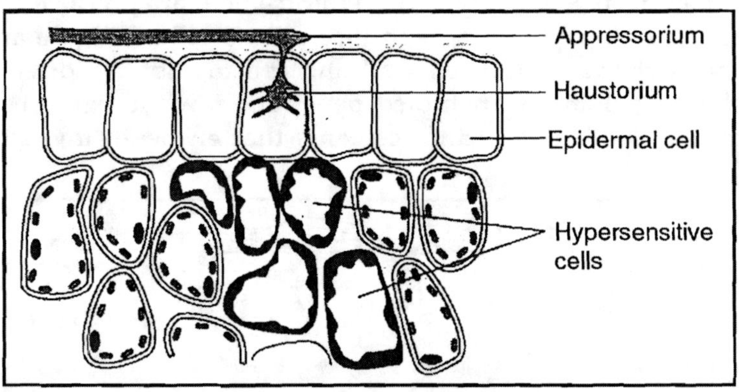

Fig. Hypersensitive Cells

This diagram shows the penetration of an epidermal cell by the appressorium of a fungal pathogen. The hypersensitive cells respond to the invasion by dying in order to prevent further spread of the pathogen.

Hypersensitive cell death is another widespread mechanism used by hosts to prevent the spread of a pathogen. Infected cells and those surrounding them "suicide", preventing further spread, and in some cases, killing the pathogen. It is often associated with the initiation of other responses, such as lignification and the synthesis of anti-microbial compounds. The success of hypersensitive cell death as a resistance mechanism depends on the nutritional requirements of the specific pathogen and the timing, magnitude and location of the host response.

Antibiotic Compounds

Phytoalexins are low molecular weight antibiotics produced by many (but not all) plants in response to infection. There are many biotic elicitors of phytoalexin production, such as cell wall components, as well as abiotic elicitors, such as heavy metals and ultraviolet light.

Phytoalexins inhibit the growth of bacteria and fungi *in vivo* and *in vitro,* and production of these antibiotics during an infection can induce resistance to subsequent infections by that pathogen. Over 350 phytoalexins are known in over 100 plant species. They include pterocarpans, sesquiterpenes, cryptophenols, isocoumarins, isoflavenoids, and others.

Phytoalexins may be produced by any part of the plant, although different phytoalexins can accumulate in different organs. Generally, related plant species produce structurally-related phytoalexins, and many produce more than one, enabling the plant to present a toxic cocktail to invading pathogens. Phytoalexins are produced in cells surrounding an infection site and delivered to the infected cell packaged in lipid vesicles, creating a toxic micro-environment in the infected cell and, hopefully, preventing disease establishment. Phytoalexin accumulation is often associated with hypersensitive cell death, although only living cells can synthesise phytoalexins. Some plants can also sequester phytoalexins into vacuoles as stores of inactive sugar-conjugates, which can be cleaved and released quickly if initial defence responses are unsuccessful.

DELAYED ACTIVE DEFENSES

Delayed active defences include containment of the

pathogen, wound repair, expression of pathogenesis-related proteins and the acquisition of systemic resistance. These mechanisms restrict the spread of the pathogen after infection is established and contain the damage to host tissues.

Physical Responses

The ability to repair wounds can help protect the plant from further infection by other, opportunistic pathogens. A secondary meristem in fleshy tissues, fruits, roots and bark, the cork cambium, can produce cork cells, which have thick, suberised walls. These cells can create a barrier to further colonisation by the pathogen and, in some cases, develop an abscission layer around the site of infection, causing the infected tissue separate from the healthy tissue.

Wounded tree trunks often secrete protective gums that seal the wound from further infection. The formation of tyloses, ingrowths if the protoplasm in xylem parenchyma can also restrict the spread of pathogenic propagules in the xylem, although they also tend to reduce the movement of water through the vessels, causing water stress in the plant.

Pathogenesis-Related Proteins

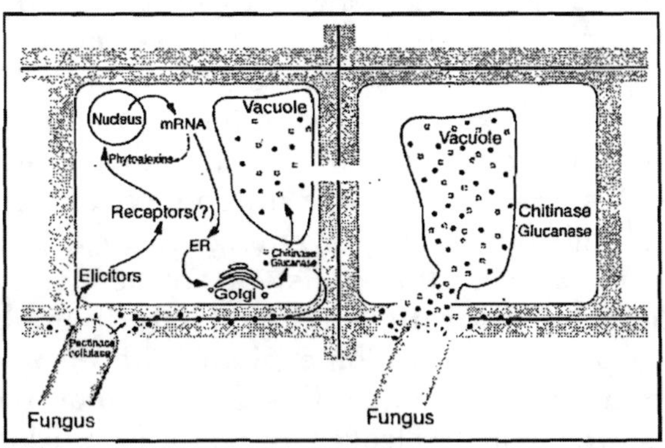

Fig. Protein Synthesis in Response to a Pathogen

This figure shows the production and movement of proteins in response to penetration of the plant cells by a fungus. Following production (left cell), chitinase and glucanase accumulate in the vacuoles, which break down, releasing their contents and disrupting pathogen acitivity.

There is a range of novel proteins synthesised in response to infection, many of which have b-glucanase, chitinase or lysozyme activity. Some pathogenesis-related proteins disrupt pathogen nutrition. The presence of low levels of these proteins in healthy plants suggests that they might have other roles in plant growth and development aside from disease resistance. Chitinase and glucanase accumulate in the vacuoles, and glucanase is also sometimes secreted into the intercellular space.

They dissolve the fungal cell wall, fragments of which then elicit hypersensitive cell death. The breakdown of the vacuole during decompartmentalisation of the cytoplasm results in a flood of hydrolytic enzymes, which have antiviral, antifungal and antibacterial activity. The accumulation of pathogenesis-related proteins peaks around 7-10 days after initial infection. The presence of these proteins *before* infection increases the plant's resistance to pathogens, as in the case of systemic acquired resistance.

SYSTEMIC ACQUIRED RESISTANCE

Systemic acquired resistance, or induced resistance, is characterised by the increased resistance of a plant to a wide range of pathogens following infection by one pathogen. It is therefore fundamentally different from the specific antigen-antibody mechanism of resistance seen in the immune response of mammals. Rather than providing immunity *per se*, systemic acquired resistance reduces the severity of later diseases.

The development of systemic acquired resistance usually requires the development of a slowly expanding necrotic lesion and other localised responses to infection, the release of a phloem-translocated signal originating from the infection site, and the subsequent priming of the plant against further attacks, allowing a more rapid response in the case of future infections.

The nature of the signal that triggers systemic acquired

resistance is as yet unknown, and is likely to be a complex signal transduction pathway mediated by a number of stress signals. Salicylic acid plays a key role, via interaction with salicylic acid-binding proteins that can cause build-up of reactive oxygen species or activate gene expression.

The levels of salicylic acid increase around necrotic lesions and remain high in plants that have acquired resistance. It is not, however, salicylic acid itself that acts as the signal that is translocated systemically throughout the plant.

Co-ordination of Defence Responses

The success of defence responses is increased if activated in combination. Passive mechanisms, coupled with rapid active responses and slower follow-up defences provide a broad defence front to the plant. The specific interaction between host and pathogen is of course crucial to the success of the plant's resistance or the pathogen's invasion, and is mediated by the many pathways involved in producing or detecting elicitors, enhancers, suppressors and secondary signals.

Chapter 5

Recognising Infectious Plant Diseases

Infectious plant diseases differ from noninfectious or abiotic disorders because they are caused by specific plant pathogens. Plant pathogens are usually microorganisms that are on or inside the plant (the host). Many plant pathogens are common in the air or soil around the host plants. Although plant pathogens may be common in many landscape situations, they cannot infect the plant and cause disease unless the environment is favorable for the interaction to take place.

Thus, in order to have an infectious plant disease, a susceptible host plant, a pathogen and a favorable environment for disease development are required. These three elements make up what is referred to as the disease triangle.

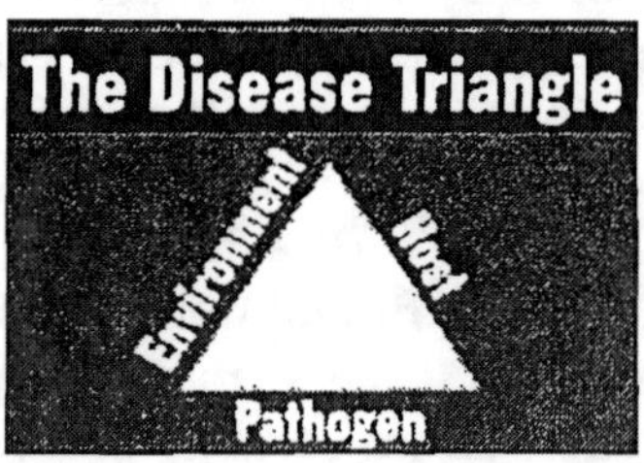

Understanding this basic principle about infectious plant disease is important to developing successful disease management programs. Many of the cultural practices that are useful in disease management are directed toward changing or creating an environment that is less conducive to disease development. Another important difference between infectious and noninfectious disease is that infectious diseases can spread from plant to plant by means of pathogen dispersal. Noninfectious diseases do not spread in this manner.

TYPES OF PLANT PATHOGENS

The most common plant pathogens on ornamental plants are fungi, bacteria, viruses, and nematodes.

FUNGI

Fungi are by far the most common type of plant pathogens. Plant pathogenic fungi cause many different symptoms, including leaf spots, wilts, curled leaves, dieback, enlargements or galls, and stunted or dead plants. Some fungi are visible in forms—called signs—such as rusts, mildews, and sooty molds.

Fungi are usually composed of fine threadlike structures (hyphae) that form a network or mass (mycelium) growing on or through their host. Fungi can spread through rhizomorphs, sclerotia, or by microscopic spores. Rhizomorphs are rootlike or cordlike masses of hyphae that can contaminate soil and plant parts. Sclerotia are compact masses of hyphae that can persist for relatively long periods in soil.

Fungal spores (which are produced in great numbers) can be spread by wind, water, soil movement, machinery, insects, or other things with which they come in contact, including people. Sclerotia, rhizomorphs, and some types of spores may survive for long periods in or on plants or in soil.

When and where conditions such as temperature, moisture, and the presence of a host plant are suitable for growth, they produce new fungal hyphae and infect the plant. If they are large enough to be seen with the naked eye or a hand lens, signs such as mycelium, masses of spores, and spore-forming structures such as mushrooms, help in identifying fungi.

BACTERIA

Bacteria are microscopic, one-celled organisms that obtain

their food from living plants or other organic matter. Common symptoms of disease resulting from bacterial infection are shoot blight, leaf spots, soft rots, scabs, wilts, cankers, and galls on branches, twigs, stems, and roots. Unlike fungi, plant pathogenic bacteria generally do not produce spores that can survive adverse environmental conditions; they must usually remain in contact with a host plant or plant debris to survive.

Plant-infecting bacteria generally require warmth and moisture to multiply, and are usually not a problem during dry summer weather, except where there is overhead irrigation. Bacteria are commonly spread by splashing water, but may also be dispersed by insects or by moving infested plants, soil, or equipment. In some instances, such as diseases of herbaceous annual flowers, bacteria may be seed-borne.

VIRUSES

Virusesare submicroscopic particles that can infect plants and lead to stunting, discoloring, deformation, or death of leaves, stems, fruit, or entire plants. Viruses rarely kill woody plants, and some infected plants exhibit no symptoms. Viruses require a living host cell in which to reproduce, and generally do not survive for very long outside of living tissue.

Many viruses are spread by aphids, leafhoppers, or other plant-feeding insects. Some viruses are spread by nematodes, budding or grafting, or the movement of infected seeds, plants, plant parts, or infested equipment. Once a plant is infected by a virus, it usually remains infected during its entire life.

NEMATODES

Nematodes are tiny (usually microscopic) roundworms that feed on a wide diversity of organisms. Nematode species that damage plants may feed on or in roots, tubers, bulbs, leaves, or stems.

With the exception of the pinewood nematode, which feeds inside branches and trunks, most species that are pests of ornamental plants feed on roots. These include root knot nematodes, cyst nematodes, root lesion nematodes, and dagger, pin, ring and stunt nematodes.

TYPES OF INFECTIOUS PLANT DISEASES OF ORNAMENTALS

POWDERY MILDEWS

Powdery mildew diseases are caused by fungi that are often specific to the host they infect. Generally, each plant type is unique in that the powdery mildew fungus that infects it will not infect any other. Powdery mildew is common on roses, lilacs, English oak, sycamore, some deciduous azaleas and zinnias, and many other plants.

The white growth on leaves and stems is the powdery mildew fungus growing on the surface of the plant tissue. Much of this growth consists of spores, which can be blown about to susceptible plants nearby. Small structures, called haustoria, grow within host cells, injuring them as they obtain food. Powdery mildew will not usually kill a plant but may weaken it and reduce winter hardiness. The unsightly fungal growth greatly reduces the aesthetic quality of the plant. If the infection is widespread, the entire leaf blade may be covered with a thin, white fungal growth. Some mildew fungi affect older leaves first, such as on lilacs. Others affect newer shoots, such as on roses or crabapples. When new shoots are affected, leaf curling and shoot stunting and twisting may be severe.

Powdery mildew can be treated effectively with chemicals. The white appearance may remain, however, even though the fungus may be dead. For highly susceptible plants such as roses, a regular preventive spray programme should be planned and carried out. For other plants, watering only early in the day and pruning out surrounding vegetation to improve sunlight and drying of plant foliage may be all that is needed for control.

RUST DISEASES

Rust diseases are also caused by fungi that are highly specific to the host plant they infect. Rust fungi usually sporulate on plant tissues forming masses of orange to dark red spores. Orange to reddish-brown spots or lesions will form on leaves, twigs, or fruit. Sometimes galling and disfigurement will be associated with the infections. In summer, infections on some plants will produce cup-like structures or tubular growths that will bear abundant orange spores. Commonly affected trees are hawthorns, crabapples, or junipers. Some kinds of rust fungi must have two different species (hosts) of plants in the area before they become a problem.

An example is Cedar-Apple rust, where both a rosaceous host (crabapple, hawthorn) and junipers are required for disease development. Some other common rust diseases include rust on aster, hollyhock, and snapdragon.

LEAF SPOTTING AND BLIGHTING

Most of these diseases are also caused by fungi; however, bacteria can also be the cause. Common bacterial diseases include bacterial leaf spot of English ivy and zinnia; fireblight on crabapple, mountain ash, hawthorn, cotoneaster, and pyracantha; crown gall on rose and euonymus; and bacterial leaf scorch of shade trees.

The fungi that cause leaf spotting and blight are common and are usually spread by microscopic spores either long distances by air or shorter distances by splashing water. Most notable among this group are Entomosporium leaf spot on hawthorn, scab on crabapple and pyracantha, Botrytis blights, black spot of rose, and anthracnose on shade trees.

Anthracnose diseases are common and often result in irregular tan or brown areas on the leaves, especially along major veins of the leaf blade. The condition usually is noticed in late spring on lower parts of the plant. Deformation and twisting of leaves often results from infections that occurred during bud break and leaf expansion.

Anthracnose blight is commonly found on ash, dogwood, white oak, maple, sycamore, and London plane tree and others, particularly when such plants are growing in damp, cool places. Frost injury, herbicide damage, and leaf scorch are often mistaken for anthracnose infection. Examine affected leaves closely for fungal fruiting bodies within or near the edge of browned areas and on the veins.

Anthracnose is also common on walnut, causing early leaf fall later in the summer. On river birch, small circular leaf spots are the symptoms. On some hosts, especially sycamore and flowering dogwood, twig blight and cankers may occur.

ROOT AND CROWN ROTS CAUSED BY WATER MOLDS

Pythium and *Phytophthora* fungi are often called water molds because they have a spore stage that is adapted to spread by swimming in water. These organisms attack a wide variety of plants, causing root rots, stem rots, and cutting rots. Many times they do not kill a plant.

Instead, they may "prune" the root system, resulting in poor growth, yellowing, or stunting of the top portion of the plant. However, under conditions of poor soil drainage (wet soil) these fungi can girdle the crown or kill major roots resulting in plant death.

These fungi are found in almost all soils. Environmental control can be achieved by improving the drainage of the soil. The fungi do not survive well in well-drained planting beds.

ROOT AND CROWN ROTS CAUSED BY OTHER FUNGI

Aside from water molds, many other fungi cause root and crown rots. The fungus *Rhizoctonia* lives in the soil and attacks a wide variety of crops. Some species of *Fusarium, Sclerotium, Cylindrocladium, Sclerotinia* and *Thielaviopsis* behave similarly. All of these fungi can persist in soil for many months through specially adapted resting structures.

CANKERS AND TWIG BLIGHTS

Many fungi such as *Botryosphaeria, Phomopsis, Sphaeropsis* (formerly *Diplodia*), and *Fusicoccum* cause stem or twig cankers. A canker is an infection of the bark and outer vascular tissue of a tree or shrub. The infections vary in colour from the surrounding healthy bark. Infected tissue may be sunken or swollen.

When cankers girdle the twig or branch, the end of the branch dies. Small black or brown bumps, spore-forming bodies of the fungus, are often seen in the canker. These may produce an ooze of orange, white, salmon, or tan spores during rainy weather. *Phytophthora*, mentioned under root rots caused by water molds, sometimes causes cankers as well as twig and stem blight.

Most notable examples include *Phomopsis* twig blight on juniper, *Dothiorella* branch blight on Shumard oak, *Sphaeropsis* tip blight on red or Austrian pine, and *Volutella* blight and canker on pachysandra. These pathogens sporulate primarily in the spring. The spores are spread by splashing water, insects, and wind.

PLANT-WILTING FUNGI

Verticillium is a fungus that causes wilting of a large number of woody ornamentals and some herbaceous plants such as maples, redbud, Russian olive, and chrysanthemum. The fungus invades injured roots, grows into the stem and plugs the vascular system of the plant. Wilting, yellowing, and browning of foliage may be followed by premature defoliation.

One limb of the plant may be affected first. Gradual dieback may be apparent. Yellow-brown, brown, black, or greenish-black streaks may be found in the outer rings of wood of infected branches, or in the stems of herbaceous plants. These streaks differentiate Verticillium wilt from most other types of plant wilting and dieback. The fungus that causes Verticillium wilt has specialized "resting" structures that allow it to persist in soil for many years.

Ophiostoma ulmi (formerly *Ceratocystis ulmi*) is the plant-wilting fungus that causes Dutch elm disease in a variety of elms. It is spread from plant to plant by root grafts or by elm bark beetles feeding on small twigs or by root grafts.

Chapter 6

Insecticidal Materials and Practices

Timeliness, thoroughness, and persistence are the watchwords of success. It is easier to keep an enemy away than to drive him away. The worst foes are often the smallest ones; and the injury is often done before they are detected. Be ready; begin early.

The results secured from the use of an insecticide or fungicide depends upon the operator.

GENERAL PRACTICES

CLEANLINESS

Much can be done to check the ravages of insects by destroying their breeding-places and hiding-places. Weeds, rubbish, and refuse should be eliminated.

Hand-picking is often still the best means of destroying insects, despite all the perfection of machinery and of materials. This is particularly true about the home grounds and in the garden. The cultivator should not scorn this method.

PROMOTING GROWTH

Any course that tends to promote vigour will be helpful in enabling plants to withstand the attacks of plant-lice and other insects.

BURNING

Larvae which live or feed in webs, like the tent-caterpillar and fall web-worm, may be burned with a torch.

The lamp or torch used in campaign parades finds its most efficient use here.

BANDING

To prevent the ascent of canker-worm moths and gypsy-moth caterpillars, various forms of sticky bands are in use. For this purpose there is no better substance than Tree Tanglefoot. It may be applied directly to the tree-trunk, but when so used leaves an unsightly mark and requires more material than when the following method is used:

- Place a strip of cotton batting three inches wide around the trunk;
- Cover this with a strip of tarred paper five inches wide;
- Draw the paper tight and fasten at the lap only with three or four tacks.
- Spread the tanglefoot on the upper two-thirds of the paper, and comb it from time to time to keep the surface sticky.

Burlap bands are made by tying or tacking a strip of burlap around the trunk and letting the edges hang down. The larvae will hide under the loose edge, where they may be killed. Banding is now little used for the codlin-moth, since spraying with poison has been found so much more effective.

FUMIGATION

Fumigating or "smoking" or "smudging" in greenhouses is performed by the slow burning of tobacco-stems. Best results are obtained when a sheet iron vessel made for the purpose is used, having holes in the bottom to supply draft.

A quart of live coals is placed in the bottom of the vessel, and about a pailful of tobacco-stems is laid on them. The stems should not blaze, but burn with a slow smudge. If they are slightly damp, better results are obtained. Some plants are injured by a very heavy smoke, and in order to avoid this injury, and also to more effectually destroy the insects, it is better to smoke rather lightly and often.

It is always well to smoke on two consecutive days, for

the insects which persist through the first "treatment, being weak, will be killed by the second. If the plants are wet, the smoke is more likely to scorch them. The smudge often injures flowers, as those of roses and chrysanthemums.In order to avoid this injury, the flowers should be covered with paper bags.

Tobacco fumes can be conveniently generated by burning strips of prepared nicotine paper, or by vaporizing a concentrated aqueous solution of nicotine in pans over alcohol or special kerosene lamps.

FUMIGATION WITH HYDROCYANIC ACID GAS

Hydrocyanic acid gas is a deadly poison, and the greatest care is required in its use. Always use 98 to 100 per cent pure potassium cyanide and a good grade of commercial sulfuric acid. The chemicals are always combined in the following proportion: Potassium cyanide, 1 ounce; sulfuric acid, 1 fluid ounce; water, 3 fluid ounces.

Always use an earthen dish, ourin the water first, and add the sulfuric acid to it. Put the required amount of cyanide in a thin paper bag, and when all is ready, drop it into the liquid and leave the room immediately. For mills and dwellings, use one ounce of cyanide for every 100 cubic feet of space. Make the doors and windows as tight as possible by placing strips of paper over the cracks. Remove the silverware and food, and if brass and nickel work cannot be removed, cover with vaseline. Place the proper amount of the acid and water for every room in two-gallon jars. Use two or more in large rooms or halls. Weigh out the potassium cyanide in paper bags, and place them near the jars. When all is ready, drop the cyanide into the jars, beginning on the top floors, since the fumes are lighter than air. In large buildings, it is frequently necessary to suspend the bags of cyanide over the jars by cords running through screw-eyes and all leading to a place near the door. By cutting all the cords at once, the cyani.le will be lowered into the jars and the operator may escape without injury. Let the fumigation continue all night, locking all outside doors, and place danger signs on the house.

FUMIGATION OF GREENHOUSES

No general formula can be given for fumigating the different kinds of plants grown in greenhouses, as the species and varieties differ greatly in their ability to withstand the effects of the gas.Ferns and roses are very susceptible to injury, and fumigation, if attempted at all, should be performed with great caution. Fumigation will not kill insect eggs, and thus must be repeated when the new brood appears. Fumigate only on dark nights when there is no wind. Have the house as dry as possible, and the temperature as near 60° as practicable.

FUMIGATION OF DORMANT NURSERY STOCK

Dormant nursery stock may be fumigated in a tight box or fumigating house made especially for the purpose. Fumigating houses are built of two thicknesses of matched boards with building paper between, and are provided with a tight-fitting door and ventilators.

The stock should be reasonably dry to avoid injury, and should be piled loosely in the house to permit a free circulation of the gas. Use one ounce of potassium cyanide for each 100 cubic feet of space, and let the fumigation continue forty minutes to one hour.

FUMIGATION OF CITRUS TREES

In this case the tree is covered with an octagonal sheet tent made of 61/2 ounce special drill or 8 ounce special army duck, and the gas is generated in the ordinary way beneath it. The tent is so marked that when in position it is an easy matter to determine the distance over the tent and the circumference at the ground.

The top line of numbers, beginning at 16 and continuing to 78, represents the distance in feet around the bottom of the tent. The outer vertical columns of larger numbers running from 10 to 59 represent the distance in feet over the top of the tent. The number of ounces of cyanide to use for a tree of known dimensions is found in that square where the vertical column headed by the distance around the tree intersects the horizontal line of figures corresponding to the distance over.

USING FUNGOUS DISEASES AS INSECTICIDES

Fungous diseases have been successfully employed against the citrus white-fly in Florida. There are seven species of fungus which attack the white-fly, and nearly all are more or less valuable. The object is to introduce some form of the fungus into orchards and on trees where it is absent.

This may be accomplished by spraying the under side of the leaves with a mixture of fungus spores in water. The mixture may be made by placing two or three fungus-bearing leaves in a quart of water, and stirring occasionally for fifteen minutes. Strain through cheesecloth, and apply to those parts of the tree most badly infested with the white-fly larvae.

The fungus may be introduced by pinning a dozen or so fungus-bearing leaves to the under side of the leaves of the tree infested with white-fly. The fungus-covered surface should face downward.

INSECTICIDAL SUBSTANCES

ARSENIC

Known to chemists as arsenious acid, or white oxide of arsenic. It is considered an unsafe insecticide, as its colour allows it to be mistaken for other substances; but in its various compounds it forms our best insecticides. From one to two grains, or less, usually prove fatal to an adult; 30 grains will usually kill a horse, 10 grains a cow; and 1 grain, or less, is usually fatal to a dog.

In case of poisoning, while awaiting the arrival of a physician, give emetics, and, after free vomiting, give milk and eggs. Sugar and magnesia in milk is useful.

A cheap and effective insecticide may be prepared from white arsenic by the following methods:—

For use with bordeaux mixture only. Sal soda, 2 pounds; water, 1 gallon; arsenic, 1 pound. Mix the white arsenic into a paste, and then add the sal soda and water, and boil until dissolved. Add water to replace any that has boiled away, so that one gallon of stock solution is the result. Use one quart of this stock solution to 50 gallons of bordeaux mixture for fruit-trees. Make sure there is enough lime in the bordeaux mixture to prevent the caustic action of the arsenic.

For use without bordeaux mixture. Sal soda, 1 pound; water, 1 gallon; white arsenic, 1 pound; quicklime, 2 pounds. Dissolve the white arsenic with the water and sal soda as above, and use this solution while hot to slake the 2 pounds of lime.

Add enough water to make 2 gallons. Use 2 quarts of this stock solution in 50 gallons of water. Arsenicals. - A term popularly used for compounds of arsenic. The leading arsenicals used in destroying insects are arsenate of lead, paris green, and london purple.

ARSENATE OF LEAD

This can be applied in a stronger mixture than other arsenical poisons, without injuring the foliage. It is, therefore, much used against beetles and other insects that are hard-to poison. It comes in the form of a paste or powder, and should be mixed thoroughly with a small amount of water before placing in the sprayer, else the nozzles will clog.

Arsenate of lead may be used with either bordeaux mixture or lime-sulfur without lessening the value of either. It is used in strengths varying from 4 to 10 pounds per 100 gallons, depending on the kind of insect to be killed.

LONDON PURPLE

An arsenite of lime, obtained as a by-product in the manufacture of aniline dyes. The composition is variable. The amount of arsenic varies from 30 to 50 per cent. The two following analyses show its composition:

- Arsenic, 43.65 per cent; rose aniline, 12.46; lime, 12.82; insoluble residue, 14.57; iron oxide, 1.16; and water, 2.27.
- Arsenic, 55.35 per cent; lime, 26.23; sulfuric acid, 0.22; carbonic acid, 0.27; moisture, 5.29. It is a finer powder than paris green, and therefore remains longer in suspension in water. It is used in the same manner as paris green, but is sometimes found to be

more caustic on foliage. This injury is due to the presence of much soluble arsenic; but it can be averted by the use of lime, as advised under paris green.

PARIS GREEN

An aceto-arsenite of copper. When pure it contains about 58 per cent of arsenic. By the provisions of the federal insecticide act of 1910, paris green must contain at least 50 per cent of arsenious oxide, and must not contain arsenic in water-soluble form equivalent to more than 31/2 per cent of arsenious oxide. It is applied in either a wet or dry condition; but in any case, it must be much diluted. For making a dry mixture, plaster, flour, air-slaked lime, road dust, or sifted wood ashes may be used. The strength of the mixture depends upon the plants and insects to which it is to be applied.

The strongest dry mixture now recommended is one part of poison to fifty of the diluent; but if the mixing is very thoroughly done, 1 part to 100, or even 200, is sufficient. Paris green is practically insoluble in water. When mixed with water, the mixture must be kept in a constant state of agitation, else the poison will settle, and the liquid from the bottom of the cask will be so strong as to do serious damage, while that from the top will be useless.

For potatoes, apple-trees, and most species of shade trees, 1 pound of poison to 200 or 250 gallons of water is a good mixture. Paris green is very likely to burn the foliage of stone fruits, especially peaches and Japanese plums, and has been generally replaced by arsenate of lead for such purposes. In all cases, the liquid should be applied with force, in a very fine spray. At some seasons of the year foliage is more liable to injury than at others. The addition of a little lime (twice the bulk of lime as of paris green) to the mixture will tend to prevent any caustic injury upon the foliage.

ARSENICALS AND FUNGICIDES

Arsenicals may be used in connection with some fungicides, and both insects and plant diseases in this manner may be combated at the same time.

The arsenicals may be added to bordeaux mixture in the same proportion as if the bordeaux were plain water. Arsenate of lead may be added to the lime-sulfur wash, but the addition of paris green or arsenite of lime is liable to cause burning.

The addition of lime to paris green and london purple mixtures greatly lessens injury to foliage, and, as a consequence, they can be applied several times stronger than ordinarily used, if they are combined with the bordeaux mixture. The free lime in the mixture combines with the soluble arsenic, which is the material that injures the foliage, and the combination is thus made quite harmless.

BAIT

Vegetable Bait

Spray a patch of clover or some other plant that the insects will eat with paris green or some other arsenical; mow it close to the ground, and while fresh place it in small piles around the infested plants. To avoid wilting of the bait, cover the heaps with a shingle or piece of board.

Bran-Arsenic Mash

White arsenic, 1/2 pound, or paris green, 1 pound; bran, 50 pounds. Mix thoroughly and then add enough water to make a wet mash.

Sugar or molasses may be added, but is unnecessary. Poisoned baits are used against cutworms and grasshoppers.

Bisulfid of Carbon

A thin liquid which volatilizes at a very low temperature, the vapour being very destructive to animal life. It is exceedingly inflammable, and should never be used near a lamp or fire. It is used for many root-insects. It is poured into a hole, which is immediately closed up, causing the fumes to permeate the soil in all directions. In loose soils it is very destructive to insects. Against weevils infesting stored grain and corn, carbon bisulfid is effective at the rate of 5 pounds for each 1000 cubic feet, provided the application is made while the temperature is not below 65° F.

Make the bins as tight as possible, and after sprinkling the liquid over the grain, cover tightly with gas-proof tarpaulin. Let the fumigation continue for at least twenty-four hours.

Carbolic Acid and Soap Mixtures

One ounce crude carbolic acid; 1 pound fish-oil soap; 1 gallon hot water. Mix thoroughly. This wash is used for borers. Apply with a cloth or soft broom. Use only on dormant wood.

Carbolic Acid Emulsion

Soap, 1 pound; water, 1 gallon; crude carbolic acid (90 per cent strength), 1 pint. Dissolve the soap in hot water; add the carbolic acid, and agitate into an emulsion. For use against root-maggots, dilute with 30 parts of water.

Criddle Mixture

Mix 1 pound of paris green with 1/2 barrel of horse droppings, and add 1 pound of salt if the material is not fresh. For use against grasshoppers.

Distillate Emulsion

5 gallons of 28° gravity untreated distillate; 5 gallons boiling water, 11/2 pounds whale-oil soap. Dissolve the soap in hot water, add the distillate, and thoroughly emulsify by means of a power pump until a yellowish, creamy emulsion is produced. For use on lemon dilute with 12 parts of water; on orange, with 15 parts. Formerly much used on citrus trees, but now generally replaced by fumigation.

Hot Water

Submerge affected plants or branches in water at a temperature of about 125°. For aphis. It will also kill rose-bugs at a temperature of 125°-135°.

Kerosene Emulsion

Hard, soft, or whale-oil soap, 1/2 pound; water, 1 gallon; kerosene, 2 gallons. Dissolve the soap in hot water; remove from the fire and while still hot add the kerosene. Pump the liquid back into itself for five or ten minutes, or until it becomes a creamy mass. If properly made, the oil will not separate out on cooling.

For use on dormant trees, dilute with from 5 to 7 parts of water. For killing plant-lice on foliage dilute with from 10 to 15 parts of water. Crude oil emulsion is made in the same way by substituting crude oil in place of kerosene. The strength of oil emulsions is frequently indicated by the percentage of oil in the diluted liquid:— For a 10 per cent emulsion add 17 gallons of water to 3 gallons stock emulsion.

For a 15 per cent emulsion add 101/3 gallons of water to 3 gallons stock emulsion.

For a 20 per cent emulsion add 7 gallons of water to 3 gallons stock emulsion.

For a 25 per cent emulsion add 5 gallons of water to 3 gallons stock emulsion.

Lime-Sulfur

A compound of lime and sulfur makes both a good insecticide and a good fungicide. There are several forms of it, as:

- The ordinary dilute home-made;
- The concentrated home-made;
- The commercial concentrated brands;
- The so-called self-boiled preparation.

The three first are solutions, and are modifications of one preparation; the self-boiled is mostly a mechanical mixture of the lime and sulfur.

- ***Home-made dilute solution of lime-sulfur:*** Quick lime, 20 pounds; sulfur (flour or flowers), 15 pounds; water, 50 gallons. The lime and sulfur must be thoroughly boiled. An iron kettle is often convenient for the work. Proceed as follows: Place the lime in the kettle. Add hot water gradually in sufficient quan-tity to produce the most rapid slaking of the lime. When the lime begins to slake, add the sulfur and stir together. If convenient, keep the mixture covered with burlap to save the heat.

After slaking has ceased, add more water, and boil the mixture one hour. As the sulfur goes into solution, a rich orange-red or dark green colour will appear.

After boiling sufficiently, add water to the required amount and strain into the spray tank. The wash is most effective when applied warm, but may be applied cold. If one has access to a steam boiler, boiling with steam is more convenient and satisfactory. Barrels may be used for holding the mixture, and the steam applied by running a pipe or rubber hose into the mixture. Proceed in the same manner as for boiling in the kettle until the lime is slaked, when the steam may be turned on. Continue boiling for forty-five minutes to an hour, or more if necessary to get the sulfur well dissolved.

This mixture can be applied safely only when the trees are dormant, — late in the autumn after the leaves have fallen, or early in the spring before the buds swell. It is mainly an insecticide for San Jose scale, although it has considerable value as a fungicide for certain diseases, like the peach leaf-curl. As the San Jose scale is not killed unless the solution comes in contact with it, great care should be exercised to completely cover the branches.

Home-Made Concentrated Lime-Sulfur Wash

For making the concentrated mixture, the steps are the same as in making the usual boiled wash, but the following formula should be used: —

Lump lime { Pure calcium oxide........................36 1b.
or
95 per ct. calcium oxide..................38 1b
or
90 per ct. calcium oxide..................40 1b }

Sulfur...80 1b.

Water...50 gal.

Slake the lime, make a thin paste, and add the sulfur. Flowers of sulfur or light or heavy sulfur flour may be used. The lime should be fresh lump lime, free from dirt and grit, containing 90 per cent or more of calcium oxide and less than 5 per cent of magnesium oxide. Stir thoroughly during the hour of cooking, to break up the lumps of sulfur. Enough water should be added at the start so that the evaporation will not leave the quantity less than 50 gallons when the cooking is ended.

If kettles are used, 10 to 15 gallons additional will be needed, while with steam none may be required. The kettles should be considerably larger than the amount of wash to be made, to prevent loss of material by boiling over. The clear liquid should be drawn off into tight containers if to be kept any considerable time; and stored where there is no danger of temperatures much below freezing. For use, test the clear solution with the hydrometer, and dilute as indicated in the table: —

Dilutions of Concentrated Lime-Sulfur Solutions for Spraying:

DEGREES BEAUME	SPECIFIC GRAVITY	WITH EACH GALLON OF CONCENTRATE, USE —		DEGREES BEAUME	SPECIFIC GRAVITY	WITH EACH GALLON OF CONCENTRATE, USE —	
		For San Jose scale	For Blister Mite			For San Jose scale	For Blister Mite
		Gals. water	Gals. water			Gals. water	Gals. water
35	1.3181	9	12	25	1.2083	5½	7½
34	1.3063	8½	11½	24	1.1983	5¼	7
33	1.2946	8	11	23	1.1885	4¾	6¾
32	1.2831	7 2/4	10½	22	1.1788	4½	6¼
31	1.2719	7½	10	21	1.1693	4¼	5¾
30	1.2608	7	9½	20	1.1600	4	5½
29	1.2500	6¾	9	19	1.1507	3⅗	5
28	1.2393	6½	8¾	18	1.1417	3⅓	4¾
27	1.2288	6	8¼	17	1.1328	3	4½
26	1.2184	5¾	7¾	16	1.1240	2⅘	4
				15	1.1153	2½	3¾

Commercial Concentrated Mixtures.

The lime-sulfur may be purchased in the concentrated form and the trouble of making it avoided. The strength of the commercial product varies considerably, and in order to compute the proper dilution correctly the strength should be determined by means of a hydrometer. Having determined the strength of the concentrated mixture, the proper dilution for use against the San Jose scale and blister mite.

MISCIBLE OILS

There are now on the market a number of preparations of petroleum and other oils intended primarily for use against the San Jose scale. They mix readily with cold water, and are immediately ready for use. While quickly prepared, easily applied, and generally effective, they cost considerably more than lime-sulfur wash. They are, however, less corrosive to the pumps, and more agreeable to use.

They are especially valuable to the man with only a few trees or shrubs who would not care to go to the trouble and expense to make up the lime-sulfur wash. They should be diluted with not more than 10 or 12 parts of water. Use only on dormant trees, and when the temperature is above freezing and the trees are not wet.

PYRETHRUM

A very fine and light brown powder, made from the flower-heads of species of pyrethrum. It is scarcely injurious to man.

Three brands are on the market:

- Persian Insect-powder,
- Made from the heads of Pyrethrum roseum,
- A species also cultivated as an ornamental plant.

The plant is native to the Caucasus region. Delmation Insect-powder, made from Pyrethrum cinerariaefolium. Buhach, made in California from cultivated plants of Pyrethrum cinerariaefolium.

When fresh and pure, all these brands appear to be equally valuable, but the home-grown product is usually considered most reliable. Pyrethrum soon loses its value when exposed to the air. It is used in various ways: —

- In solution in water, 1 ounce to 3 gallons. Should be mixed up twenty-four hours before using.
- Dry, without dilution. In this form it is excellent for thrips and lice on roses and other bushes. Apply when the bush is wet. Useful for aphis on house plants.

- Dry, diluted with flour or any light and fine powder. The poison may be used in the proportion of 1 part to from 6 to 30 of the diluent.
- In fumigation. It may be scattered directly upon coals, or made into small balls by wetting and molding with the hands and then set upon coals. This is a desirable way of dealing with mosquitoes and flies.
- In alcohol.
  - Put a part of pyrethrum (buhach) and 4 parts alcohol, by weight, in any tight vessel. Shake occasionally, and after eight days filter. Apply with an atomizer. Excellent for greenhouse pests. For some plants it needs to be diluted a little.
  - Dissolve about 4 ounces of powder in 1 gill of alcohol, and add 12 gallons of water.
- Decoction. Whole flower-heads are treated to boiling water, and the liquid is covered to prevent evaporation. Boiling the liquid destroys its value.

Good insect-powder can be made from Pyrethrum roseum, and probably also from P. cinerariaefolium, grown in the home garden.

RESIN AND FISH-OIL COMPOUND

Ten pounds of resin; 11/2 pounds of fish-oil, 3 pounds of caustic soda, and enough water to make 50 gallons.

Break the resin into small lumps, and place it together with the caustic soda in the boiler, with three or four inches of water.

Stir till the resin is dissolved; then add about one-fourth of the required water and boil one-half hour. Place in the spray tank and add the rest of the water.

Used in California against the cottony cushion scale and the brown apricot scale.

SOAPS, WHALE-OIL, OR FISH-OIL SOAP.

Soaps are effective insecticides for plant-lice. Dissolve in hot water and dilute so as to obtain 1 pound of soap for every

5 or 7 gallons of water. Commercial whale-oil or fish-oil soaps frequently injure tender foliage because of the free alkali which they contain.

An excellent fish-oil soap free from uncombined alkali may be easily prepared at home, as follows:

- Six pounds of caustic soda
- 1/2 gallons of water
- 22 pounds of fish oil.

Completely dissolve the caustic soda in the water, and then add the fish-oil very gradually, under constant and vigorous stirring. The combination occurs readily at ordinary summer temperatures and boiling is unnecessary. Stir briskly for about twenty minutes after the last of the oil has been added.

Soap and Tobacco

Dissolve 8 pounds of the best soft soap in 12 gallons of rain-water, and when cold add 1 gallon of strong tobacco liquor. For plant-lice.

Soda and Aloes

Dissolve 2 pounds of washing-soda and 1 ounce of bitter Barbadoes aloes, and when cold add one gallon of water. Dip the plants into the solution, and lay them on their sides for a short time, and the insects will drop off. Syringe the plants with clean, tepid water, and return to the house. For plant-lice.

Sulfur: Fumes of sulfur are destructive to insects, but should be carefully used, or plants will be injured. The sulfur should be evaporated over an oil stove, until the room is filled with the vapour.

The sulfur should never be burned, as burning sulfur kills plants. For greenhouse use.

Sulfur and Water

To 3 gallons of weak soap suds add 1 pound of flowers of sulfur and stir thoroughly. Apply as a spray. For red spider and mites.

Tanglefoot is a sticky commercial substance much used for banding trees.

Tar is sometimes used to prevent the female and wingless canker-worm from ascending trees.

The tar should be placed on cotton, or some material which will prevent it from coming in contact with the bark, and a band of the preparation is then placed around the trunk. Care must be taken to see that the tar does not injure the tree.

Tarred paper may be rolled loosely about trees to keep away mice, but it should be removed before warm weather. It is sometimes recommended as a preventive of the attacks of borers, but it very often injures trees, and should be used, if at all, with great caution.

TOBACCO

- Stems, placed on the walks and under the benches of greenhouses, for plant-lice. Renew it every month.
- Tobacco-water, used with whale-oil soap.
- Dust and snuff. Snuff may be blown lightly on plants, as house-plants, for lice.
- Fumes. Burn dampened tobacco-stems.
- Nicotyl. Steep tobacco-stems in water, and evaporate the water.
- Tea, or common decoction. Boil the stems or dust thoroughly, and strain. Then add cold water until the decoction contains 2 gallons of liquid to 1 pound of tobacco.

There are various concentrated commercial preparations of tobacco which have recently been giving good results against plant-lice.

WHITE HELLEBORE

A light brown powder made from the roots of the white hellebore plant (Veratrum album), one of the lily family. It is applied both dry and in water.

In the dry state, it is usually applied without dilution, although the addition of a little flour will render it more adhesive. In water, 4 ounces of the poison is mixed with 2 or 3 gallons; and an ounce of glue, or thin flour paste, is sometimes added to make it adhere.

A decoction is made by using boiling water in the same proportions. Hellebore soon loses its strength, and a fresh article should always be demanded.

It is much less poisonous than the arsenicals, and should be used in place of them upon ripening fruit. Used for various leaf-eating insects, particularly for the currant-worm and rose-slug.

Chapter 7

Treatment of Injurious Insects

Insects are of two kinds as respects their manner of taking food, the mandibulate insects, or those that chew or bite their food, as larvae (" worms ") and most beetles; and those that suck their food, as the plant-lice and true bugs. The former class is dispatched by poisons, the latter by caustic applications, as kerosene or soap preparations.

UNCLASSIFIED PESTS

Angleworm or Earthworm: The common angleworm often destroys greenhouse plants by its burrowing. It is sometimes annoying in gardens also.

Treatment: Lime-water applied to the soil.

Aphides, Plant-lice or Green-fly, and Bark-lice: Minute insects of various kinds, feeding upon the tender parts of many plants, both indoors and out.

Treatment: Kerosene emulsion. Hot water (about 125°). Pyrethrum. Fish-oil soap. Tobacco-water or extracts. Alcoholic and water extracts of pyrethrum. Hughes' fir-tree oil. In the greenhouse, fumigation with tobacco or hydrocyanic acid gas. Knock them off with the hose. In window gardens, dry pyrethrum or snuff.

Bag-worm or Basket-worm (Thyridopteryx ephemeraeformis): Larva working in singular dependent bags, and feeding upon many kinds of trees, both evergreen and deciduous. In winter the bags, empty or containing eggs, are conspicuous, hanging from the branches.

Treatment: Hand-picking. Arsenicals.

Blister-beetle (Lytta, two or three species): Soft-shelled, long-necked and slim black or gray spry beetles, feeding on the leaves of many trees and garden plants.

Treatment: Arsenicals.

Brown-tail moth (Euproctis chrysorrhaea): This highly destructive European insect was introduced near Boston a number of years ago, and is now rapidly spreading over New England. The snow-white moths, with a large tuft of brown hairs at the tip of the abdomen, appear in July and deposit eggs on the leaves in elongate masses covered with brown hairs from the body of the female.

The caterpillars become only partly grown the first season, and hibernate in conspicuous nests, three or four inches long, at the tips of the branches. The black-bodied caterpillars, clothed with rather long, brownish, stinging hairs, complete their growth the next spring, feeding ravenously on the tender foliage and causing great damage in orchards, parks, and forests.

Treatment: Cut out and burn all winter nests before the buds start. In the spring spray with arsenate of lead, as recommended for the gipsy-moth. Prevent the ascent of caterpillars from other trees by banding the trunks with tanglefoot. Keep the bands fresh by combing the surface every few days.

Cutworm: Various species of Agrotis and related genera. Soft brown or gray worms, of various kinds, feeding on the roots, crown, or even the tops of plants.

Treatment: Encircle the stem of the plant with heavy paper or tin, coating the top with tanglefoot. Arsenicals sprinkled upon small bunches of fresh grass or clover, which are scattered at short intervals about the garden towards evening. They will often collect under boards or blocks.

Arsenicals mixed with shorts and placed about the plants. Make two or three deep holes by the side of the plant with a pointed stick; the worms will fall in and cannot escape. Dig them out. Plow infested land in the fall to give birds a chance to find the worms.

Cutworm, Climbing: Several species. The worms climb

grape vines and small trees of various kinds at night and eat out the buds.

Preventive: Band of cotton batting tied about the tree by lower edge, and the top rolled down like a boot-leg.

Treatment: Arsenicals. Hellebore.

Flea-beetle (Phyllotreta vittata; Haltica striolata, etc.): Minute, dark-colored beetles, feeding upon many plants, as turnip, cabbage, radish, mustard, potato, strawberry, and stocks. They jump upon being disturbed. Closely related species attack various plants. Very destructive to plants which are just appearing above the surface.

Treatment: Bordeaux mixture applied liberally is the best remedy, it drives them away.

Four-striped Plant-bug (Paecilocapsus lineatus): A bright yellow; black-striped bug about one-third of an inch long, puncturing the young leaves and shoots of many plants.

Treatment: Jarring at any time of day into a dish of dilute kerosene. Kerosene emulsion (diluted five times) when the bugs are young, in their nymphal stage. Cut off and burn the tips of the growing shoots in early spring to destroy the eggs.

GALLS

Gipsy-moth (Porthetria dispar): Larva, nearly two inches long when mature, very hairy, nearly black, with a yellow stripe along back and sides. Devours many kinds of foliage. It has become a serious pest.

Treatment: Spray with arsenate of lead as soon as the caterpillars hatch in the spring. Band trees with tanglefoot.

May-beetle or May-bug (Lachnosterna fusca): A large and familiar brown beetle, feeding upon the leaves of many kinds of trees.

The common white grub is the larval state. It often does great damage to sod and to strawberries. Sometimes called June-bug.

Mealy-bug (Pseudococcus citri and P. longifilis): A white, scale-like insect, attacking greenhouse plants.

Treatment: Whale-oil soap. Carbolic acid and soap. Removing insects with brush on tender plants. House-plants may be washed in soapsuds. The best procedure in greenhouses is to knock them off with the hose. A small, hard stream of water upsets their domestic affairs.

Nematode Root-gall (Heterodera radicicola): A disease characterized by the knotting and contortion of the roots of the peach, orange, and many other plants. The knots are mostly rather soft swellings, and on the smaller roots. It is usually most destructive on the peach. It is caused by a nematode, or true worm (not an insect). Gulf States. Attacks greenhouse plants in the North.

Preventive: Plant non-infested plants in fresh soil; bud into healthy stocks. Fertilize highly, particularly with potassic fertilizers. Set the trees 8 or 10 inches deep in high and dry soils. Infested small trees may be remedied, in part at least, by transplanting them into highly manured holes which have been prepared contiguous to them. Does not live in regions where the ground freezes deeply. If it is feared in greenhouses, see that the soil has been thoroughly frozen before it is used. Whitewash the benches.

Red-spider or mite (Tetranychus bimaculatus): A small mite infesting many plants, both in the greenhouse and out of doors. It flourishes in dry atmospheres, and on the under sides of the leaves. In some forms it is reddish, but usually light-colored and two-spotted. Common.

Remedies: Persistent syringing with water will generally destroy them, if the spray is applied to the under surface. Use lots of force and little water to avoid drenching the beds. Sulfur and water. Dry sulfur. Sulphocide.

San Jose Scale (Aspidiotus perniciosus): This scale is nearly circular in outline and about the size of a pin-head. When abundant it forms a crust on the branches, and causes small red spots on the fruit. It multiplies with marvelous rapidity, there being three or four broods annually, and each mother scale may give birth to several hundred young.

The young are born alive, and breeding continues until late autumn, when all stages are killed by the cold weather, except the tiny, half-grown, black scales, many of which hibernate safely.

Spray thoroughly in the fall after the leaves drop, or early in the spring before growth begins, with lime-sulfur wash, or miscible oil, 1 gallon in 10 gallons of water. When badly infested, make two applications, one in the fall and another in the spring. In case of large, old trees, 25 per cent crude oil emulsion should be applied just as the buds are swelling.

Scale-insects: Various species of small insects inhabiting the young growth of trees, and sometimes the fruit, in one stage characterized by a stationary scale-like appearance. Lime-sulfur and miscible oils are the best remedies. Species which migrate on to the young growth in spring can be readily dispatched at that time by kerosene emulsion.

Snails: These animals are often very troublesome in greenhouses, eating many plants voraciously.

Preventives: Trap them by placing pieces of turnip, cabbage, or potatoes about the house. Scatter bits of camphor-gum about the plants. Strew a line of salt along the edges of the bed. Lime dusted about the plants will keep them away.

White ants or termites: These insects often infest orchard trees in the southern states, particularly in orchards which contain old stumps or rubbish.

Remedy: The soap-and-arsenites wash brushed over the trunk and branches of the tree.

Wire-worm (various species): Slim and brown larvae, feeding upon the roots of various plants. They are the larvae of the click-beetle, or snapping-beetle.

Remedy: Arsenicals sprinkled upon baits of fresh clover or other material which is placed about the field under blocks or boards. Sweetened corn-meal dough also makes a good bait. The best treatment is to plow infested land early in the fall. A system of short rotations of crops will lessen injury from wire-worms.

Chapter 8

Biological Control of Plant Pests

Biological control of plant pests and pathogens continues to inspire research and development in many fields. Plant pathogens are just one class of targets of biological control, which also is designed to limit other pests such as insects, parasitic nematodes, and weeds. In the narrowest sense, biocontrol involves suppressing pest organisms with other organisms. However, the interrelationships of many environmental variables can result in multiple interactions among organisms and their environment, several of which might contribute to effective biological control.

Furthermore, natural products and chemical compounds discovered as a result of basic research into the molecular mechanisms of pathogenesis and biological control have led to the development of "biorational" pesticides. Here, we use the term biological control in the broader sense as we describe the current status of research, commercial development, and application of biocontrol strategies targeted at plant pathogens. We conclude by describing future prospects for using biological control to limit the damage of plant pathogens in both conventional and organic agriculture.

RESEARCH

Begonias were grown in the greenhouse and inoculated with *Botrytis cinerea* under conditions optimal for the development of disease. Treatments differing in their efficacy are shown, from left to right: untreated (Un), $CaCl_2$, chlorothalonil (Fungi), and the biocontrol agent *Trichoderma hamatum* T382 inoculated into the potting mix (T382). Over the past one hundred years, research has repeatedly demonstrated that phylogenetically diverse microorganisms can act as natural antagonists of various plant pathogens. The interactions between microorganisms and plant hosts can be complex. Interactions that lead to biocontrol can include antibiosis, competition, induction of host resistance, and predation.

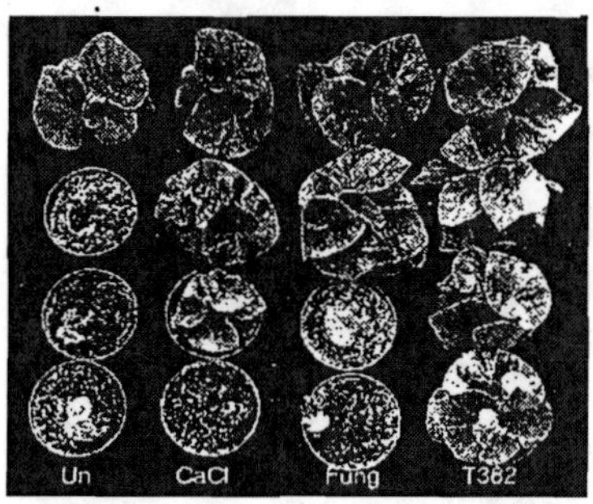

Fig. Example Bioassay for Biological Control of a Plant Pathogen

When testing bacterial and fungal isolates from the environment for biocontrol activities, between 1 and 10 per cent show at least some capacity to inhibit the growth of pathogens *in vitro*. However, fewer isolates can suppress plant diseases under diverse growing conditions and fewer still have broad-spectrum activity against multiple pathogenic taxa.

Nonetheless, intensive screens have yielded numerous candidate organisms for commercial development. Some of the microbial taxa that have been successfully commercialized and are currently marketed as EPA-registered biopesticides in the United States include bacteria belonging to the genera *Agrobacterium, Bacillus, Pseudomonas,* and *Streptomyces* and fungi belonging to the genera *Ampelomyces, Candida, Coniothyrium,* and *Trichoderma*. Screening is a critical step in the development of biocontrol agents. The success of all subsequent stages depends on the ability of a screening procedure to identify an appropriate candidate.

Many useful bacterial biocontrol agents have been found by observing zones of inhibition in Petri plates. However, this method does not identify biocontrol agents with other modes of action such as parasitism, induced plant resistance, or some forms of competition. Screening methods for parasitism include burying and retrieving propagules of the pathogen to isolate parasites. For competition, methods include looking for microbes that quickly colonize sterilized soil and have the ability to exclude other organisms that attempt to invade the space, and looking for microbes that colonize the infection court. Primary screens for new biocontrol microbes are still undertaken, and it seems likely that continued prospecting will be required to diversify the potential applications of biocontrol as well as replace more widely used biocontrol products should resistance develop.

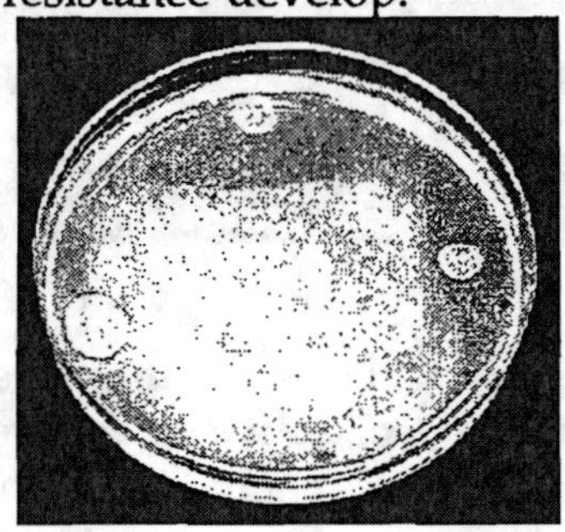

Fig. Example in *Vitro* Iinhibition Assay

Different bacterial isolates are tested for their ability to inhibit the growth of *Pythium* spp. SF2, a soilborne plant pathogen that can cause damping-off of corn and soybean seedlings. Zones of growth inhibition can be detected around strains placed at two of the four positions on the plate

In addition, other areas of research in the field of plant pathology have opened up new opportunities for disease control. Research on the mechanisms of biocontrol employed by effective bacterial strains has revealed a variety of natural products that can be exploited for the development of chemical control measures. One well-known example is pyrrolnitrin, a natural product produced by some *Pseudomonas* spp. That compound provided the chemical model for development of fludioxonil, a broad-spectrum fungicide used as seed treatment, foliar spray, or soil drench. Research into the mechanism by which plants resist bacterial pathogens led to the discovery of harpin, a protein that is now being used to activate crop defenses prior to pathogen attack. Indeed, a variety of pathogenic and non-pathogenic microorganisms can induce plant defenses and may be useful as biocontrol agents.

Other research, on the effects of organic amendments, suggests that both chemical and biological components of compost-amended soils can contribute to disease suppression. Such discoveries point to the substantial potential of diverse programs in basic research to lead to improvements in various biological control strategies.

A variety of research questions remain to be fully answered about the nature of biological control and the means to most effectively manage it under production conditions. Advanced molecular techniques are now being used to characterize the diversity, abundance, and activities of microbes that live in and around plants, including those that significantly impact plant health. Still, much remains to be learned about the microbial ecology of both plant pathogens and their microbial antagonists in different agricultural systems. Fundamental work remains to be done on characterizing the different mechanisms by which organic amendments reduce plant disease. More studies on the practical aspects of mass-production and formulation need to be undertaken to make new biocontrol products stable, effective, safer and more cost-effective.

COMMERCIAL DEVELOPMENT

Interest in biological control research continues reflecting the desire of multiple constituencies to develop sustainable methods for controlling plant disease. Growing concerns about environmental health and safety have led to substantial regulatory changes in the past several years.

The FQPA reconciles previous discrepancies between pesticide regulation under the Federal Insecticide, Fungicide, and Rodenticide Act (FIFRA) and the Federal Food Drug, and Cosmetic Act (FFDCA). Such concerns have led to increased restrictions on a variety of chemical pesticides, including some of those used to suppress plant diseases. The IR4 programme also encourages the development of biorational pesticides.

Fig. Composite Picture of Several Products Currently Being Sold as Biopesticides for the Control of Plant Diseases

The CREES Biological Control Working Group, which brings together representatives of state and federal agencies and the private sector, have published a white paper defining a new paradigm for biocontrol and outlining action steps to foster further development of the industry.

In addition, the majority of growers continue to express interest in biologically-based pest management strategies of all types as central components of an IPM approach. Such market realities promote the development of biocontrol products. Still, the path to developing and applying effective biocontrol methods is a long one, fraught with a variety of difficulties. Scientific, regulatory, business management, and marketing issues must all be handled effectively for a biocontrol product to be successful in the marketplace.

Currently in the US, a number of biologically-based products are being sold for the control of plant diseases. A growing number of companies are also developing new products that are in the process of being registered. Many of these companies are small, privately owned firms with a limited product-line (e.g., Galltrol produced by AgBiochem).

Others are publicly traded and have substantial capitalization values of between 10 and 100 million dollars (e.g., Eden Bioscience).

In addition, larger, billion-dollar companies with more diverse product lines that include a variety of agricultural chemicals and biotechnologies (e.g., Gufstafson LLC, Monsanto, Syngenta) have played a significant role in the development and marketing of products for the control of plant pathogens. Because the cost of registering a product with the EPA as a biopesticide can be between $200,000 and $500,000, a minimum capitalization close to 1 million dollars is required for establishing a start-up company.

The time to complete registration currently averages under 18 months, but the data generation requirements for different types of products and the level of experience a company has with past registrations can significantly impact the time to registration. To assist companies with the registration process, the EPA sponsors workshops each year. Biocontrol products are either marketed as stand-alone products or formulated as mixtures with other microbials (e.g., High Stick N/T) or synthetic fungicides (e.g., Kodiak).

Some products with biocontrol properties may not be registered, but are sold instead as plant strengtheners or growth promoters without any specific claims regarding disease control. To help improve the global market perception of biopesticides as effective products, the Biopesticide Industry Alliance is establishing a certification process to ensure industry standards for efficacy, quality, and consistency (P. Marrone, *personal communication*).

Fig. Electronic Monitoring of Yield and Moisture at Harvest

Successful application of biological controls requires more knowledge-intensive management. To this end, extension personnel and growers must also become more fully aware of the costs and benefits that biopesticides can provide.

Understanding when and where biological control of plant pathogens can be profitable requires an appreciation of its place within integrated pest management systems. For this purpose, the model of a pyramid can be used to highlight the way growers might productively construct their pest management programs.

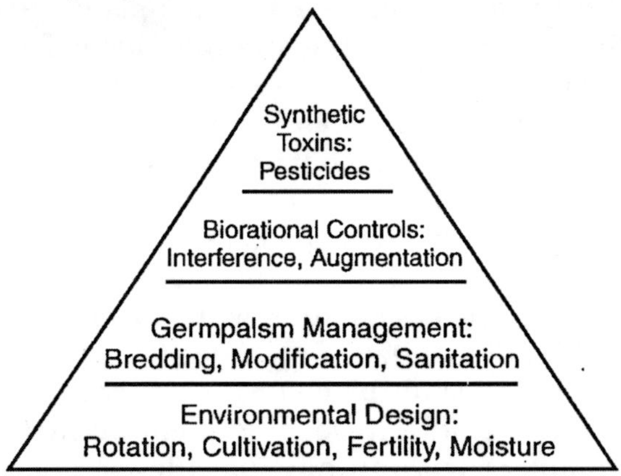

Fig. General Model for an Integrated Pest Management (IPM) Programme

In this model, the foundation of a sound pest and disease management programme in an annual cropping system begins with sound cultural practices that alter the farm landscape to promote crop health. These include crop rotations that limit the availability of host material used by plant pathogens. Judicious use of tillage can disrupt pest and pathogen life cycles, bury weeds, and prepare seed beds of optimal moisture and bulk density.

Careful management of soil fertility and moisture can also limit plant disease by minimizing plant stress. In nurseries and greenhouses, environmental control can be more tightly regulated in terms of temperature, light, moisture, and soil composition (i.e., peat-based potting mixes), but the design of such systems cannot wholly eliminate pest problems.

The second layer of defence against pests consists of the

quality of crop germplasm. Breeding for pathogen resistance contributes substantially to crop success in most regions. Newer technologies that directly incorporate genes into crop genomes, commonly referred to as genetic modification or genetic engineering, are bringing new traits into crop germplasm. The most-widely distributed of these plant-incorporated protectants (PIPs) are the different insecticidal proteins derived from *Bacillus thuringiensis.*

Other technologies, such as seed washing, testing for pathogens, and treatment are also used to keep germplasm pathogen-free. In perennial cropping systems, such as orchards and forests, germplasm quality may be more important than cultural practices, because rotation and tillage cannot be used as regularly. Upon these two layers, growers can further reduce pest and pathogen pressure by considering both biological and chemical inputs.

Biologically-based inputs such as pheromone traps and microbial pesticides can be used to interfere with pest activities. Registered biopesticides are generally labeled with short reentry intervals (REI) and pre-harvest intervals (PHI), giving greater flexibility to growers who need to balance their operational requirements and pest management goals. When living organisms are introduced, they may also augment natural beneficial populations to further reduce the damage caused by targeted pathogens or pests.

Finally, synthetic chemicals may be added to limit crop damage by pests and pathogens. Because of growing concerns about health and environmental safety, the use of toxic, carcinogenic, and/or environmentally damaging chemicals is currently being discouraged. Within the context of IPM programs, the use of such chemicals should be considered only after other management options have been fully implemented.

Clearly, the different levels described by this model are not entirely independent of one another. Indeed, the numerous biotic and abiotic variables that influence crop health are rarely independent of one another. Such complexities point to the need for more integrative, multidisciplinary research on agricultural systems that will provide a deeper understanding of the conditions under which microbial biocontrol agents might be most productively applied.

The goals of an IPM programme are simple minimize the economic and environmental costs of controlling plant pests and achieve sustainable agricultural production. All disease management strategies have associated costs. Growers need to carefully assess the potential costs and benefits of different disease management strategies.

Conventional growers may have more options available to them in terms of chemical alternatives, but economic pressures to limit rotations or reduce tillage may promote more serious disease problems.

In contrast, certified organic farmers are prevented from using most pesticides, including those commonly used as seed treatments, but longer and more diverse rotations may limit disease problems. In both cases though, thorough disease assessments are needed to determine the amount of disease pressure present and the causal agents.

If the severity of a particular disease problem justifies management action beyond the first two tiers of the IPM model described above, then on-site testing of biocontrol products can generally be recommended. Because research has shown that the efficacy of microbial biocontrols can vary somewhat from site to site and year to year (e.g., 2,3), trials should include the use of side-by-side comparisons with untreated controls and adequate replication across the farm over multiple growing seasons.

Table. List of Biocontrol Products Commercially Available for the Control of Plant Pathogens.

Bacteria	
Agrobacterium radiobacter	Galltrol Nogall
Bacillus spp.:	Companion HiStick N/T Kodiak Serenade YieldShield

Bacteria	
Burkholderia cepacia	Deny Intercept
Pseudomonas spp.	BioJect Spot-Less Bio-save, BlightBan Cedomon
Streptomyces spp.	Actinovate Mycostop
Fungi	
Ampelomyces quisqualis	AQ10
Candida oleophila	Aspire
Coniothyrium minitans	Contans WG/Intercept WG
Myrothecium verrucaria (killed)	DiTera
Trichoderma spp./	Plantshield/Rootshield/T-22
Gliocaldium spp.	Planter box, Soilgard Primastop
Plant Activators	
Bacteria	Serenade, YieldShield
Bacterial protein	Messenger
Synthetic chemicals	Actigard

Table. Biocontrol Products Commercially Available.

Actigard	
Biocontrol Agent	acibenzolar-S-methyl
Target Pathogen/Disease	many
Crop	tobacco, tomato, lettuce, spinach.
Formulation	water dispersible granule.
Application Method	drench
Actinovate	
Biocontrol Organism	*Streptomyces lydicus*
Target Pathogen/Disease	soilborne disease
Crop	greenhouse and nursery crops, turf
Formulation	water-dispersible granule
Application Method	drench
AQ10 Biofungicide	
Biocontrol Organism	*Ampelomyces quisqualis* isolate M-10
Target Pathogen/Disease	powdery mildew

Bacteria	
Crop	apples, cucurbits, grapes, ornamentals, strawberries, and tomatoes
Formulation	water-dispersible granule
Application Method	spray
Aspire	
Biocontrol Organism	*Candida oleophila* I-182
Target Pathogen/Crop	*Botrytis* spp., *Penicillium* spp.
Crop	citrus, pome fruit
Formulation	wettable powder
Application Method	post harvest application to fruit as drench, drip or spray
BioJect Spot-Less	
Biocontrol Agent	*Pseudomonas aureofaciens*
Target Pathogen/Disease	dollar spot, Anthracnose, *Pythium aphanidermatum*, Michrochium patch (pink snow mold)
Crop	turf and others
Formulation	liquid
Application	over-head irrigation; can only be used with the BioJect Automatic Fermentation System
Bio-save 10LP, 110	
Biocontrol Organism	*Pseudomonas syringae*
Target Pathogen/Disease	*Botrytis cinerea, Penicillium* spp., *Mucor pyroformis, Geotrichum candidum*
Crop	pome fruit, citrus, cherries, and potatoes
Formulation	lyophilized product, frozen cell concentrated pellets
Application Method	pellets added to water to produce liquid suspension, postharvest application to fruit as drench, dip or spray
BlightBan A506	
Biocontrol Organism	*Pseudomonas fluorescens* A506
Target Pathogen/Disease	frost damage, *Erwinia amylovora*, and

Bacteria	
Crop	russet-inducing bacteria almond, apple, apricot, blueberry, cherry, peach, pear, potato, strawberry, tomato
Formulation	wettable powder
Application Method	bloom time spray of the flower and fruit
Cedomon	
Biocontrol Organism	*Pseudomonas chlororaphis* strain
Target Pathogen/Disease	leaf stripe, net blotch, *Fusarium* sp., spot blotch, leaf spot, and others
Crop	barley and oats; potential for other cereals
Formulation	seed treatment
Application Method	seed dressing
Companion	
Biocontrol Organism	*Bacillus subtilis*GB03, *other B. subtilis,B. lichenformis, B. megaterium*
Target Pathogen/Disease	*Rhizoctonia, Pythium, Fusarium,*and *Phytophthora*
Crop	greenhouse and nursery
Formulation	liquid
Application Method	drench at time of seeding and transplanting or as a spray for turf
Contans WG, Intercept WG	
Biocontrol Organism	*Coniothyrium minitans*
Target Pathogen/Disease	*Sclerotinia sclerotiorum* and *S. minor*
Crop	all agricultural soils.
Formulation	water dispersible granule
Application Method	spray
Deny	
Biocontrol Organism	*Burkholderia cepacia*type Wisconsin
Target Pathogen/Disease	*Rhizoctonia, Pythium, Fusarium,* and disease caused by lesion, spiral, lance, and sting nematodes
Crop	alfalfa, barley, beans, clover, cotton, peas, grain sorghum, vegetable

Bacteria	
	crops, and wheat
Formulation	peat-based dried biomass from solid fermentation; aqueous suspension
Application Method	applied to seeds with a sticking agent in planter box (aqueous suspension formulation is for use in drip irrigation or as a seedling drench)
DiTera Biocontrol	
Organism	fermentation product of killed *Myrothecium verrucaria*
Target Pathogen/Disease	Parasitic nematodes
Crop	cole crops, grape, ornamentals, turf, trees
Formulation	powder and liquid suspension from fermentation
Application Method	applied to soil and incorporated by various methods
Galltrol	
Biocontrol Organism	*Agrobacterium radiobacter* Strain 84
Target Pathogen/Disease	crown gall disease caused by *Agrobacterium tumefaciens*
Crop	fruit, nut, and ornamental nursery stock
Formulation	petri plates with pure culture grown on agar
Application Method	bacterial mass from one plate transferred to one gallon of non-chlorinated water; suspension applied to seeds, seedlings, cuttings, roots, stems, and as soil drench
HiStick *N/T*	
Biocontrol Organism	*Bacillus subtilis* MBI600.
Target Pathogen/Disease	*Fusarium, Rhizoctonia, Aspergillus*
Crop	soybean, alfalfa, dry/snap beans, peanuts
Application method	slurry, damp and dry inoculation of

Bacteria	
	seed
Intercept	
Biocontrol Organism	*Burkholderia cepacia*
Target Pathogen/Disease	*hizoctonia solani, Fusarium* spp., *Pythium* sp.
Crop	maize, vegetables, cotton
Kodiak (several formulations)	
Biocontrol Agent	*Bacillus subtilis* GB03
Target Pathogen/Disease	*Rhizoctonia solani, Fusarium* spp., *Alternaria* spp., and *Aspergillus* spp. that attack roots
Crop	cotton, legumes
Formulation	dry powder; usually applied with chemical fungicides
Application	added to a slurry mix for seed treatment; hopper box treatment
Messenger	
Biocontrol Agent	*Erwinia amylovora* HrpN harpin protein.
Target Pathogen/Disease	many
Crop	field, ornamental, and vegetable crops
Formulation	powder
Application	drench or spray.
Mycostop	
Biocontrol Agent	*Streptomyces griseoviridis* strain K61
Target Pathogen/Disease	*Fusarium* spp., *Alternaria brassicola, Phomopsis* spp., *Botrytis* spp., *Pythium* spp., and *Phytophthora* spp. that cause seed, root, and stem rot, and wilt disease
Crop	field, ornamental, and vegetable crops
Formulation	powder
Application	drench, spray or through irrigation system
Nogall	
Biocontrol Agent	Agrobacterium radibacter K1026

Bacteria	
Target Pathogen/Disease	crown gall disease caused by *Agrobacterium tumefaciens*
Crop	fruit, nut, and ornamental nursery stock
Formulation	petri plates with pure culture grown on agar
Application Method	bacterial mass from one plate transferred to one gallon of non-chlorinated water; suspension applied to seeds, seedlings, cuttings, roots, stems, and as soil drench
Primastop	
Biocontrol Agent	*Gliocladium catenulatum*
Target Pathogen/Disease	soilborne pathogens that cause seed, root, and stem rot, and wilt disease
Crop	ornamental, vegetable, and tree crops
Formulation	powder
Application	drench, spray or through irrigation system
RootShield, Plant Shield, T-22 Planter box	
Biocontrol Agent	*Trichoderma harzianum* Rifai strain KRL-AG2 (T-22)
Target Pathogen/Disease	*Pythium* spp., *Rhizoctonia solani, Fusarium* spp.
Crop	trees, shrubs, transplants, all ornamentals, cabbage, tomato, cucumber
Formulation	granules or wettable powder
Application	granules mixed with soil or potting medium; powder mixed with water and added as a soil drench
Serenade	
Biocontrol Agent	*Bacillus subtilis* QST716.
Target Crop/Pathogen	powdery mildew, downy mildew, Cercospora leaf spot, early blight, late blight, brown rot, fire blight,

Bacteria	
	and others
Crop	cucurbits, grapes, hops, vegetables, peanuts, pome fruits, stone fruits, and others
Formulation	wettable powder
Application	spray
SoilGard	
Biocontrol Agent	*Gliocladium virens* (a.k.a. *Trichoderma virens*) GL-21
Target Crop/Pathogen	damping-off and root rot pathogens especially *Rhizoctonia solani* and *Pythium* spp.
Crop	ornamental and food crop plants grown in greenhouses, nurseries, homes, and interiorscapes
Formulation	granules
Application	granules are incorporated in soil or soilless growing media prior to seeding
YiedShield	
Biocontrol Agent	*Bacullus pumillus* GB34
Target Pathogen/Disease	soilborne fungal pathogens causing root diseases
Crop	soybean
Application	dry powder added to a slurry mix for seed treatment; hopper box treatment

Chapter 9

Biological Control of Plant Pathogens

Plant diseases need to be controlled to maintain the quality and abundance of food, feed, and fibre produced by growers around the world. Different approaches may be used to prevent, mitigate or control plant diseases. Beyond good agronomic and horticultural practices, growers often rely heavily on chemical fertilizers and pesticides.

Such inputs to agriculture have contributed significantly to the spectacular improvements in crop productivity and quality over the past 100 years. However, the environmental pollution caused by excessive use and misuse of agrochemicals, as well as fear-mongering by some opponents of pesticides, has led to considerable changes in people's attitudes towards the use of pesticides in agriculture.

Today, there are strict regulations on chemical pesticide use, and there is political pressure to remove the most hazardous chemicals from the market. Additionally, the spread of plant diseases in natural ecosystems may preclude successful application of chemicals, because of the scale to which such applications might have to be applied.

Consequently, some pest management researchers have focused their efforts on developing alternative inputs to synthetic chemicals for controlling pests and diseases. Among these alternatives are those referred to as biological controls. A variety of biological controls are available for use, but further development and effective adoption will require a greater understanding of the complex interactions among plants, people, and the environment.

BIOLOGICAL CONTROL AGENT

The terms "biological control" and its abbreviated synonym "biocontrol" have been used in different fields of biology, most notably entomology and plant pathology. In entomology, it has been used to describe the use of live predatory insects, entomopathogenic nematodes, or microbial pathogens to suppress populations of different pest insects.

In plant pathology, the term applies to the use of microbial antagonists to suppress diseases as well as the use of host specific pathogens to control weed populations. In both fields, the organism that suppresses the pest or pathogen is referred to as the biological control agent (BCA). More broadly, the term biological control also has been applied to the use of the natural products extracted or fermented from various sources.

These formulations may be very simple mixtures of natural ingredients with specific activities or complex mixtures with multiple effects on the host as well as the target pest or pathogen. And, while such inputs may mimic the activities of living organisms, nonliving inputs should more properly be referred to as biopesticides or biofertilizers, depending on the primary benefit provided to the host plant. The various definitions offered in the scientific literature have sometimes caused confusion and controversy.

For example, members of the U.S. National Research Council took into account modern biotechnological developments and referred to biological control as "the use of natural or modified organisms, genes, or gene products, to reduce the effects of undesirable organisms and to favor desirable organisms such as crops, beneficial insects, and microorganisms", but this definition spurred much subsequent debate and it was frequently considered too broad by many scientists who worked in the field.

Because the term biological control can refer to a spectrum of ideas, it is important to stipulate the breadth of the term when it is applied to the review of any particular work.

Definitions of biocontrol differ depending on the target of suppression; number, type and source of biological agents; and the degree and timing of human intervention.

Most broadly, biological control is the suppression of damaging activities of one organism by one or more other organisms, often referred to as natural enemies. With regards to plant diseases, suppression can be accomplished in many ways. If growers' activities are considered relevant, cultural practices such as the use of rotations and planting of disease resistant cultivars (whether naturally selected or genetically engineered) would be included in the definition. Because the plant host responds to numerous biological factors, both pathogenic and non-pathogenic, induced host resistance might be considered a form of biological control.

More narrowly, biological control refers to the purposeful utilization of introduced or resident living organisms, other than disease resistant host plants, to suppress the activities and populations of one or more plant pathogens. This may involve the use of microbial inoculants to suppress a single type or class of plant diseases.

Or, this may involve managing soils to promote the combined activities of native soil- and plant-associated organisms that contribute to general suppression. Most narrowly, biological control refers to the suppression of a single pathogen (or pest), by a single antagonist, in a single cropping system. Most specialists in the field would concur with one of the narrower definitions presented above. In this review, biological control will be narrowly defined as highlighted above in bold.

TYPES OF INTERACTIONS CONTRIBUTING TO BIOLOGICAL CONTROL

Throughout their lifecycle, plants and pathogens interact with a wide variety of organisms. These interactions can significantly affect plant health in various ways. In order to understand the mechanisms of biological control, it is helpful to appreciate the different ways that organisms interact. Note, too, that in order to interact, organisms must have some form of direct or indirect contact. Odum proposed that the interactions of two populations be protocooperation, commensalism, neutralism, competition, amensalism, parasitism, and predation. While the terminology was developed for macroecology, examples of all of these types of interactions can be found in the natural world at both the macroscopic and microscopic level.

And, because the development of plant diseases involves both plants and microbes, the interactions that lead to biological control take place at multiple levels of scale. From the plant's perspective, biological control can be considered a net positive result arising from a variety of specific and non-specific interactions. Using the spectrum of Odum's concepts, we can begin to classify and functionally delineate the diverse components of ecosystems that contribute to biocontrol.

Mutualism is an association between two or more species where both species derive benefit. Sometimes, it is an obligatory lifelong interaction involving close physical and biochemical contact, such as those between plants and mycorrhizal fungi. However, they are generally facultative and opportunistic. For example, bacteria in the genus *Rhizobium* can reproduce either in the soil or, to a much greater degree, through their mutualistic association with legume plants. These types of mutualism can contribute to biological control, by fortifying the plant with improved nutrition and/or by stimulating host defenses.

Protocooperation is a form of mutualism, but the organisms involved do not depend exclusively on each other for survival. Many of the microbes isolated and classified as BCAs can be considered facultative mutualists involved in protocooperation, because survival rarely depends on any specific host and disease suppression will vary depending on the prevailing environmental conditions.

Further down the spectrum, commensalism is a symbiotic interaction between two living organisms, where one organism benefits and the other is neither harmed nor benefited. Most plant-associated microbes are assumed to be commensals with regards to the host plant, because their presence, individually or in total, rarely results in overtly positive or negative consequences to the plant. And, while their presence may present a variety of challenges to an infecting pathogen, an absence of measurable decrease in pathogen infection or disease severity is indicative of commensal interactions.

Neutralism describes the biological interactions when the population density of one species has absolutely no effect whatsoever on the other. Related to biological control, an inability to associate the population dynamics of pathogen with that of another organism would indicate neutralism.

In contrast, antagonism between organisms results in a negative outcome for one or both. Competition within and between species results in decreased growth, activity and/or fecundity of the interacting organisms. Biocontrol can occur when non-pathogens compete with pathogens for nutrients in and around the host plant. Direct interactions that benefit one population at the expense of another also affect our understanding of biological control.

Parasitism is a symbiosis in which two phylogenetically unrelated organisms coexist over a prolonged period of time. In this type of association, one organism, usually the physically smaller of the two (called the parasite) benefits and the other (called the host) is harmed to some measurable extent. The activities of various hyperparasites, i.e., those agents that parasitize plant pathogens, can result in biocontrol. And, interestingly, host infection and parasitism by relatively avirulent pathogens may lead to biocontrol of more virulent pathogens through the stimulation of host defence systems.

Predation refers to the hunting and killing of one organism by another for consumption and sustenance. While the term predator typically refer to animals that feed at higher trophic levels in the macroscopic world, it has also been applied to the actions of microbes, e.g., protists, and mesofauna, e.g. fungal feeding nematodes and microarthropods, that consume pathogen biomass for sustenance.

Biological control can result in varying degrees from all of these types of interactions, depending on the environmental context within which they occur. Significant biological control, as defined above, most generally arises from manipulating mutualisms between microbes and their plant hosts or from manipulating antagonisms between microbes and pathogens.

MECHANISMS OF BIOLOGICAL CONTROL

Because biological control can result from many different types of interactions between organisms, researchers have focused on characterizing the mechanisms operating in different experimental situations. In all cases, pathogens are antagonized by the presence and activities of other organisms that they encounter.

Here, we assert that the different mechanisms of antagonism occur across a spectrum of directionality related to the amount of interspecies contact and specificity of the interactions. Direct antagonism results from physical contact and/or a high-degree of selectivity for the pathogen by the mechanism(s) expressed by the BCA(s).

Table Types of Interspecies Antagonisms Leading to Biological Control of Plant Pathogens.

Type	Mechanism	Examples
Direct antagonism	Hyperparasitism/ predation	Lytic/some nonlytic mycoviruses Ampelomyces quisqualis Lysobacter enzymogenes Pasteuria penetrans Trichoderma virens
Mixed-path antagonism	Antibiotics	2, 4-diacetylphloroglucinol Phenazines Cyclic lipopeptides
	Lytic enzymes	Chitinases Glucanases Proteases
	Unregulated waste products	Ammonia Carbon dioxide Hydrogen cyanide
	Physical/chemical interference	Blockage of soil pores Germination signals con-

Type	Mechanism	Examples
		sumption Molecular cross-talk confused
Indirect antagonism	Competition	Exudates/leachates consumption Siderophore scavenging Physical niche occupation
	Induction of host resistance	Contact with fungal cell walls Detection of pathogen-associated, molecular patterns Phytormone-mediated induction

In such a scheme, hyperparasitism by obligate parasites of a plant pathogen would be considered the most direct type of antagonism because the activities of no other organism would be required to exert a suppressive effect. In contrast, indirect antagonisms result from activities that do not involve sensing or targeting a pathogen by the BCA(s). Stimulation of plant host defence pathways by non-pathogenic BCAs is the most indirect form of antagonism.

However, in the context of the natural environment, most described mechanisms of pathogen suppression will be modulated by the relative occurrence of other organisms in addition to the pathogen. While many investigations have attempted to establish the importance of specific mechanisms of biocontrol to particular pathosystems, all of the mechanisms described below are likely to be operating to some extent in all natural and managed ecosystems.

And, the most effective BCAs studied to date appear to antagonize pathogens using multiple mechanisms. For instance, pseudomonads known to produce the antibiotic 2,4-diacetylphloroglucinol (DAPG) may also induce host defenses. Additionally, DAPG-producers can aggressively colonize roots, a trait that might further contribute to their ability to suppress pathogen activity in the rhizosphere of wheat through competition for organic nutrients.

Hyperparasites and Predation

In hyperparasitism, the pathogen is directly attacked by a specific BCA that kills it or its propagules. In general, there are four major classes of hyperparasites: obligate bacterial pathogens, hypoviruses, facultative parasites, and predators. *Pasteuria penetrans* is an obligate bacterial pathogen of root-knot nematodes that has been used as a BCA. Hypoviruses are hyperparasites. A classical example is the virus that infects *Cryphonectria parasitica*, a fungus causing chestnut blight, which causes hypovirulence, a reduction in disease-producing capacity of the pathogen. The phenomenon has controlled the chestnut blight in many places. However, the interaction of virus, fungus, tree, and environment determines the success or failure of hypovirulence. There are several fungal parasites of plant pathogens, including those that attack sclerotia (e.g. *Coniothyrium minitans*) while others attack living hyphae (e.g. *Pythium oligandrum*). And, a single fungal pathogen can be attacked by multiple hyperparasites.

For example, *Acremonium alternatum*, *Acrodontium crateriforme*, *Ampelomyces quisqualis*, *Cladosporium oxysporum*, and *Gliocladium virens* are just a few of the fungi that have the capacity to parasitize powdery mildew pathogens. Other hyperparasites attack plant-pathogenic nematodes during different stages of their life cycles (e.g. *Paecilomyces lilacinus* and *Dactylella oviparasitica*). In contrast to hyperparasitism, microbial predation is more general and pathogen non-specific and generally provides less predictable levels of disease control. Some BCAs exhibit predatory behaviour under nutrient-limited conditions. However, such activity generally is not expressed under typical growing conditions.

For example, some species of *Trichoderma* produce a range of enzymes that are directed against cell walls of fungi. However, when fresh bark is used in composts, *Trichoderma* spp. do not directly attack the plant pathogen, *Rhizoctonia solani*. But in decomposing bark, the concentration of readily available cellulose decreases and this activates the chitinase genes of *Trichoderma* spp., which in turn produce chitinase to parasitize *R. solani.*

Antibiotic-Mediated Suppression

Antibiotics are microbial toxins that can, at low concentrations, poison or kill other microorganisms. Most microbes produce and secrete one or more compounds with antibiotic activity. In some instances, antibiotics produced by microorganisms have been shown to be particularly effective at suppressing plant pathogens and the diseases they cause. In all cases, the antibiotics have been shown to be particularly effective at suppressing growth of the target pathogen *in vitro* and/or *in situ.*

Table: Some of Antibiotics Produced by BCAs

Antibiotic	Source	Target pathogen	Disease
2, 4-diacetyl-phloroglucinol	Pseudomonas fluorescens F113	Phthium spp.	Damping off
Agrocin 84	Agrobacterium radiobacter	Agrobacterium tumefaciens	Crown gall
Bacillomycin D	Bacillus subtilis AU195	Aspergillus flavus	Aflatoxin contamination
Bacillomycin, fengycin	Bacillus amyloliquefaciens FZB42	Fusarium oxysporum	Wilt
Xanthobaccin A	Lysobacter sp. strain SB-K88	Aphanomyces cochlioides	Damping off
Gliotoxin	Trichoderma virens	Rhizoctonia solani	Root rots
Herbicolin	Pantoea agglo-merans C9-1	Erwinia amylovora	Fire blight
Iturin A	B. subtilis QST713	Botrytis cinerea and R. solani	Damping off
Mycosubtilin	B. Subtilis BBG100	Pythium aphanidermatum	Damping off
Phenazines	P. fluorescens 2-79 and 3084	Gaeumannomyces graminis var. tritici	Take-all
Pyoluteorin,	P. fluorescens	Pythium ultimum	Damping off

Antibiotic	Source	Target pathogen	Disease
pyrrolnitrin		and R. solani	
Pyrrolnitrin, pseudane	Burkholderia cepacia	R. solani and Phricularia oryzae	Damping off and rice blast
Zwittermicin A	Bacillus cereus UW85	Phytophthora medicaginis and P.aphanidermatum	Damping off

To be effective, antibiotics must be produced in sufficient quantities near the pathogen to result in a biocontrol effect. *In situ* production of antibiotics by several different biocontrol agents has been measured; however, the effective quantities are difficult to estimate because of the small quantities produced relative to the other, less toxic, organic compounds present in the phytosphere.

And while methods have been developed to ascertain when and where biocontrol agents may produce antibiotics, detecting expression in the infection court is difficult because of the heterogenous distribution of plant-associated microbes and the potential sites of infection. In a few cases, the relative importance of antibiotic production by biocontrol bacteria has been demonstrated, where one or more genes responsible for biosynthesis of the antibiotics have been manipulated.

For example, mutant strains incapable of producing phenazines or phloroglucinols have been shown to be equally capable of colonizing the rhizosphere but much less capable of suppressing soilborne root diseases than the corresponding wild-type and complemented mutant strains. Several biocontrol strains are known to produce multiple antibiotics which can suppress one or more pathogens.

For example, *Bacillus cereus* strain UW85 is known to produce both zwittermycin and kanosamine. The ability to produce multiple antibiotics probably helps to suppress diverse microbial competitors, some of which are likely to be plant pathogens.

The ability to produce multiple classes of antibiotics, that differentially inhibit different pathogens, is likely to enhance biological control. More recently, *Pseudomonas putida* WCS358r strains genetically engineered to produce phenazine and DAPG displayed improved capacities to suppress plant diseases in field-grown wheat.

Lytic Enzymes and other Byproducts of Microbial Life

Diverse microorganisms secrete and excrete other metabolites that can interfere with pathogen growth and/or activities. Many microorganisms produce and release lytic enzymes that can hydrolyze a wide variety of polymeric compounds, including chitin, proteins, cellulose, hemicellulose, and DNA. Expression and secretion of these enzymes by different microbes can sometimes result in the suppression of plant pathogen activities directly. For example, control of *Sclerotium rolfsii* by *Serratia marcescens* appeared to be mediated by chitinase expression.

And, a b–1,3–glucanase contributes significantly to biocontrol activities of *Lysobacter enzymogenes* strain C3. While they may stress and/or lyse cell walls of living organisms, these enzymes generally act to decompose plant residues and nonliving organic matter. Currently, it is unclear how much of the lytic enzyme activity that can be detected in the natural environment represents specific responses to microbe-microbe interactions. It seems more likely that such activities are largely indicative of the need to degrade complex polymers in order to obtain carbon nutrition. Nonetheless, microbes that show a preference for colonizing and lysing plant pathogens might be classified as biocontrol agents.

Lysobacter and *Myxobacteria* are known to produce copious amounts of lytic enzymes, and some isolates have been shown to be effective at suppressing fungal plant pathogens. So, the lines between competition, hyperparasitism, and antibiosis are generally blurred. Furthermore, some products of lytic enzyme activity may contribute to indirect disease suppression.

For example, oligosaccharides derived from fungal cell walls are known to be potent inducers of plant host defenses. Interestingly, *Lysobacter enzymogenes* strain C_3 has been shown to induce plant host resistance to disease, though the precise activities leading to this induction are not entirely clear. The quantitative contribution of any and all of the above compounds to disease suppression is likely to be dependent on the composition and carbon to nitrogen ratio of the soil organic matter that serves as a food source for microbial populations in the soil and rhizosphere. However, such activities can be manipulated so as to result in greater disease suppression. For example, in postharvest disease control, addition of chitosan can stimulate microbial degradation of pathogens similar to that of an applied hyperparasite. Chitosan is a non-toxic and biodegradable polymer of beta-1,4-glucosamine produced from chitin by alkaline deacylation. Amendment of plant growth substratum with chitosan suppressed the root rot caused by *Fusarium oxysporum radicis-lycopersici* in tomato. Although the exact mechanism of action of chitosan is not fully understood, it has been observed that treatment with chitosan increased resistance to pathogens. Other microbial byproducts also may contribute to pathogen suppression. Hydrogen cyanide (HCN) effectively blocks the cytochrome oxidase pathway and is highly toxic to all aerobic microorganisms at picomolar concentrations.

The production of HCN by certain fluorescent pseudomonads is believed to be involved in the suppression of root pathogens. *P. fluorescens* CHAO produces antibiotics, siderophores and HCN, but suppression of black rot of tobacco caused by *Thielaviopsis basicola* appeared to be due primarily to HCN production. Howell reported that volatile compounds such as ammonia produced by *Enterobacter cloacae* were involved in the suppression of *Pythium ultimum*-induced damping-off of cotton. While it is clear that biocontrol microbes can release many different compounds into their surrounding environment, the types and amounts produced in natural systems in the presence and absence of plant disease have not been well documented and this remains a frontier for discovery.

COMPETITION BETWEEN PATHOGENS AND NON-PATHOGENS

From a microbial perspective, soils and living plant surfaces are frequently nutrient limited environments. To successfully colonize the phytosphere, a microbe must effectively compete for the available nutrients. On plant surfaces, host-supplied nutrients include exudates, leachates, or senesced tissue.

Additionally, nutrients can be obtained from waste products of other organisms such as insects (e.g. aphid honeydew on leaf surface) and the soil. While difficult to prove directly, much indirect evidence suggests that competition between pathogens and non-pathogens for nutrient resources is important for limiting disease incidence and severity.

In general, soilborne pathogens, such as species of *Fusarium* and *Pythium,* that infect through mycelial contact are more susceptible to competition from other soil- and plant-associated microbes than those pathogens that germinate directly on plant surfaces and infect through appressoria and infection pegs.

Genetic work of Anderson revealed that production of a particular plant glycoprotein called agglutinin was correlated with potential of *P. putida* to colonize the root system. *P. putida* mutants deficient in this ability exhibited reduced capacity to colonize the rhizosphere and a corresponding reduction in Fusarium wilt suppression in cucumber.

The most abundant nonpathogenic plant-associated microbes are generally thought to protect the plant by rapid colonization and thereby exhausting the limited available substrates so that none are available for pathogens to grow. For example, effective catabolism of nutrients in the spermosphere has been identified as a mechanism contributing to the suppression of *Pythium ultimum* by *Enterobacter cloacae.* At the same time, these microbes produce metabolites that suppress pathogens.

These microbes colonize the sites where water and carboncontaining nutrients are most readily available, such as exit points of secondary roots, damaged epidermal cells, and nectaries and utilize the root mucilage. Biocon ιol based on competition for rare but essential micronutrients, such as iron, has also been examined. Iron is extremely limited in the rhizosphere, depending on soil pH.

In highly oxidized and aerated soil, iron is present in ferric form, which is insoluble in water (pH 7.4) and the concentration may be as low as 10^{-18} M. This concentration is too low to support the growth of microorganisms, which generally need concentrations approaching 10^{-6} M. To survive in such an environment, organisms were found to secrete iron-binding ligands called siderophores having high affinity to sequester iron from the micro-environment.

Almost all microorganisms produce siderophores, of either the catechol type or hydroxamate type. Kloepper were the first to demonstrate the importance of siderophore production as a mechanism of biological control of *Erwinia carotovora* by several plant-growthpromoting *Pseudomonas fluorescens* strains A1, BK1, TL3B1 and B10. And, a direct correlation was established *in vitro* between siderophore synthesis in fluorescent pseudomonads and their capacity to inhibit germination of chlamydospores of *F. oxysporum*.

As with the antibiotics, mutants incapable of producing some siderophores, such as pyoverdine, were reduced in their capacity to suppress different plant pathogens. The increased efficiency in iron uptake of the commensal microorganisms is thought to be a contributing factor to their ability to aggressively colonize plant roots and an aid to the displacement of the deleterious organisms from potential sites of infection.

Induction of Host Resistance

Plants actively respond to a variety of environmental stimuli, including gravity, light, temperature, physical stress, water and nutrient availability. Plants also respond to a variety of chemical stimuli produced by soil- and plant-associated microbes. Such stimuli can either induce or condition plant host defenses through biochemical changes that enhance resistance against subsequent infection by a variety of pathogens.

Induction of Host Defenses

Induction of host defenses can be local and/or systemic

in nature, depending on the type, source, and amount of stimuli. Recently, phytopathologists have begun to characterize the determinants and pathways of induced resistance stimulated by biological control agents and other non-pathogenic microbes.

The first of these pathways, termed systemic acquired resistance (SAR), is mediated by salicylic acid (SA), a compound which is frequently produced following pathogen infection and typically leads to the expression of pathogenesis-related (PR) proteins. These PR proteins include a variety of enzymes some of which may act directly to lyse invading cells, reinforce cell wall boundaries to resist infections, or induce localized cell death.

A second phenotype, first referred to as induced systemic resistance (ISR), is mediated by jasmonic acid (JA) and/or ethylene, which are produced following applications of some nonpathogenic rhizobacteria. Interestingly, the SA- and JA-dependent defence pathways can be mutually antagonistic, and some bacterial pathogens take advantage of this to overcome the SAR. For example, pathogenic strains of *Pseudomonas syringae* produce coronatine, which is similar to JA, to overcome the SA-mediated pathway.

Table. Bacterial Determinants and types of Host Resistance Induced by Biocontrol Agents

Bacterial strain	Plant species	Bacterial determinant	Type
Bacillus mycoides strain Bac J	Sugar beet	Peroxidase, chitinase and β-1,3-glucanase	ISR
Bacillus pumilus 203-6	Sugar beet and β-1,3-glucanase	Peroxidase, chitinase	ISR
Bacillus subtilis GB03 and IN937	Arabidopsis	2,3-butanediol	ISR
Pseudomonas fluorescens strains CHA0	Tobacco	Siderophore	SAR
	Arabidopsis	Antibiotics (DAPG)	ISR
WCS374	Radish	Lipopolysaccharide	
		Siderophore	

Bacterial strain	Plant species	Bacterial determinant	Type
		Iron regulated factor	
WCS417	Carnation	Lipopolysaccharide	ISR
	Radish	Lipopolysaccharide	ISR
		Iron regulated factor	
	Arabidopsis	Lipopolysaccharide	ISR
	Tomato	Lipopolysaccharide	ISR
Pseudomonas	Arabidopsis	Lipopolysaccharide	ISR
WCS 358	Arabidopsis	Lipopolysaccharide	ISR
		Siderophore	ISR
BTP1	Bean	Z,3-hexenal	ISR
Serratia marcescens 90-166	Cucumber	Siderophore	ISR

Because the various host-resistance pathways can be activated to varying degrees by different microbes and insect feeding, it is plausible that multiple stimuli are constantly being received and processed by the plant. Thus, the magnitude and duration of host defence induction will likely vary over time. Only if induction can be controlled, i.e. by overwhelming or synergistically interacting with endogenous signals, will host resistance be increased.

A number of strains of root-colonizing microbes have been identified as potential elicitors of plant host defenses. Some biocontrol strains of *Pseudomonas* sp. and *Trichoderma* sp. are known to strongly induce plant host defenses. In several instances, inoculations with plant-growth-promoting rhizobacteria (PGPR) were effective in controlling multiple diseases caused by different pathogens, including anthracnose (*Colletotrichum lagenarium*), angular leaf spot (*Pseudomonas syringae* pv. *lachrymans* and bacterial wilt (*Erwinia tracheiphila*). A number of chemical elicitors of SAR and ISR may be produced by the PGPR strains upon inoculation, including salicylic acid, siderophore, lipopolysaccharides, and 2,3-butanediol, and other volatile substances.

Again, there may be multiple functions to such molecules blurring the lines between direct and indirect antagonisms. More generally, a substantial number of microbial products have been identified as elicitors of host defenses, indicating that host defenses are likely stimulated continually over the course of a plant's lifecycle. Excluding the components directly related to pathogenesis, these inducers include lipopolysaccharides and flagellin from Gram-negative bacteria; cold shock proteins of diverse bacteria; transglutaminase, elicitins, and glucans in Oomycetes; invertase in yeast; chitin and ergosterol in all fungi; and xylanase in *Trichoderma*.

These data suggest that plants would detect the composition of their plant-associated microbial communities and respond to changes in the abundance, types, and localization of many different signals. The importance of such interactions is indicated by the fact that further induction of host resistance pathways, by chemical and microbiological inducers, is not always effective at improving plant health or productivity in the field.

MICROBIAL DIVERSITY AND DISEASE SUPPRESSION

Plants are surrounded by diverse types of mesofauna and microbial organisms, some of which can contribute to biological control of plant diseases. Microbes that contribute most to disease control are most likely those that could be classified competitive saprophytes, facultative plant symbionts and facultative hyperparasites. These can generally survive on dead plant material, but they are able to colonize and express biocontrol activities while growing on plant tissues.

A few, like avirulent *Fusarium oxysporum* and binucleate *Rhizoctonia*-like fungi, are phylogenetically very similar to plant pathogens but lack active virulence determinants for many of the plant hosts from which they can be recovered. Others, like *Pythium oligandrum* are currently classified as distinct species. However, most are phylogenetically distinct from pathogens and, most often, they are subspecies variants of the same microbial groups.

Due to the ease with which they can be cultured, most biocontrol research has focused on a limited number of bacterial (*Bacillus, Burkholderia, Lysobacter, Pantoea, Pseudomonas,* and *Streptomyces*) and fungal (*Ampelomyces, Coniothyrium, Dactylella, Gliocladium, Paecilomyces,* and

Trichoderma) genera. Still, other microbes that are more recalcitrant to *in vitro* culturing have been intensively studied. These include mycorrhizal fungi, e.g. *Pisolithus* and *Glomus* spp. that can limit subsequent infections, and some hyperparasites of plant pathogens, e.g. *Pasteuria penetrans* which attack root-knot nematodes. Because multiple infections can and do take place in fieldgrown plants, weakly virulent pathogens can contribute to the suppression of more virulent pathogens, via the induction of host defenses.

Lastly, there are the many general micro- and meso-fauna predators, such as protists, collembola, mites, nematodes, annelids, and insect larvae whose activities can reduce pathogen biomass, but may also facilitate infection and/or stimulate plant host defenses by virtue of their own herbivorous activities. While various epiphytes and endophytes may contribute to biological control, the ubiquity of mycorrhizae deserves special consideration.

Mycorrhizae

Mycorrhizae are formed as the result of mutualist symbioses between fungi and plants and occur on most plant species. Because they are formed early in the development of the plants, they represent nearly ubiquitous root colonists that assist plants with the uptake of nutrients (especially phosphorus and micronutrients).

The vesicular arbuscular mycorrhizal fungi (VAM, also known as arbuscular mycorrhizal or endomycorrhizal fungi) are all members of the zygomycota and the current classification contains one order, the Glomales, encompassing six genera into which 149 species have been classified.

Arbuscular mycorrhizae involve aseptate fungi and are named for characteristic structures like arbuscles and vesicles found in the root cortex. Arbuscules start to form by repeated dichotomous branching of fungal hyphae approximately two days after root penetration inside the root cortical cell. Arbuscules are believed to be the site of communication between the host and the fungus. Vesicles are basically hyphal swellings in the root cortex that contain lipids and cytoplasm intra- and inter- cellular and can often develop thick walls in older roots. These thick walled structures may function as propagules.

During colonization, VAM fungi can prevent root infections by reducing the access sites and stimulating host defence. VAM fungi have been found to reduce the incidence of root-knot nematode. Various mechanisms also allow VAM fungi to increase a plant's stress tolerance. This includes the intricate network of fungal hyphae around the roots which block pathogen infections.

Inoculation of apple-tree seedlings with the VAM fungi *Glomus fasciculatum* and *G. macrocarpum* suppressed apple replant disease caused by phytotoxic myxomycetes. VAM fungi protect the host plant against root-infecting pathogenic bacteria. The damage due to *Pseudomonas syringae* on tomato may be significantly reduced when the plants are well colonized by mycorrhizae.

The mechanisms involved in these interactions include physical protection, chemical interactions and indirect effects. The other mechanisms employed by VAM fungi to indirectly suppress plant pathogens include enhanced nutrition to plants; morphological changes in the root by increased lignification; changes in the chemical composition of the plant tissues like antifungal chitinase, isoflavonoids; alleviation of abiotic stress and changes in the microbial composition in the mycorrhizosphere.

In contrast to VAM fungi, ectomycorrhizae proliferate outside the root surface and form a sheath around the root by the combination of mass of root and hyphae called a mantle. Disease protection by ectomycorrhizal fungi may involve multiple mechanisms including antibiosis, synthesis of fungistatic compounds by plant roots in response to mycorrhizal infection and a physical barrier of the fungal mantle around the plant root.

Ectomycorrhizal fungi like *Paxillus involutus* effectively controlled root rot caused by *Fusarium oxysporum* and *Fusarium moniliforme* in red pine. Inoculation of sand pine with *Pisolithus tinctorius*, another ectomycorrhizal fungus, controlled disease caused by *Phytophthora cinnamomi*. Because plant diseases may be suppressed by the activities of one or more plantassociated microbes, researchers have attempted to characterize the organisms involved in biological control. Historically, this has been done primarily through isolation, characterization, and application of individual organisms. By design, this approach focuses on specific forms of disease suppression.

Specific Suppression

Specific suppression results from the activities of one or just a few microbial antagonists. This type of suppression is thought to be occurring when inoculation of a biocontrol agent results in substantial levels of disease suppressiveness. Its occurrence in natural systems may also occur from time to time.

For example, the introduction of *Pseudomonas fluorescens* that produce the antibiotic 2,4-diacetylphloroglucinol can result in the suppression of various soilborne pathogens. However, specific agents must compete with other soil- and root-associated microbes to survive, propagate, and express their antagonistic potential during those times when the targeted pathogens pose an active threat to plant health. In contrast, general suppression is more frequently invoked to explain the reduced incidence or severity of plant diseases because the activities of multiple organisms can contribute to a reduction in disease pressure.

High soil organic matter supports a large and diverse mass of microbes resulting in the availability of fewer ecological niches for which a pathogen competes. The extent of general suppression will vary substantially depending on the quantity and quality of organic matter present in a soil. Functional redundancy within different microbial communities allows for rapid depletion of the available soil nutrient pool under a large variety of conditions, before the pathogens can utilize them to proliferate and cause disease.

For example, diverse seed-colonizing bacteria can consume nutrients that are released into the soil during germination thereby suppressing pathogen germination and growth. Manipulation of agricultural systems, through additions of composts, green manures and cover crops is aimed at improving endogenous levels of general suppression.

Biocontrol Research, Development, and Adoption

Biological control really developed as an academic discipline during the 1970s and is now a mature science supported in both the public and private sector. Research related to biological control is published in many different scientific journals, particularly those related to plant pathology and entomology. Additionally, three academic journals are specifically devoted to the discipline.

In the United States, research funds for the discipline are provided primarily by several USDA programs. These include the Section 406 programs, regional IPM grants, Integrated Organic Programme, IR-4, and several programs funded as part of the National Research Initiative. Monies also exist to stimulate the development of commercial ventures through the small business innovation research (SBIR) programs. Such ventures are intended to be conduits for academic research that can be used to develop new companies.

Much has been learned from the biological control research conducted over the past forty years. But, in addition to learning the lessons of the past, biocontrol researchers need to look forward to define new and different questions, the answers to which will help facilitate new biocontrol technologies and applications. Currently, fundamental advances in computing, molecular biology, analytical chemistry, and statistics have led to new research aimed at characterizing the structure and functions of biocontrol agents, pathogens, and host plants at the molecular, cellular, organismal, and ecological levels.. Some of the research questions that will advance our understanding of biological controls and the conditions under which it can be most fruitfully applied.

Some current topics of biocontrol research and development and assoriated questions:

- The ecology of plant-associated microbes

  - How are pathogens and their antagonists distributed in the environment?
  - Under what conditions do biocontrol agents exert their suppressive capacities?
  - How do native and introduced populations respond to different management practices?
  - What determines successful colonization and expression of biocontrol traits?
  - What are the components and dynamics of plant host defence induction?
- Application of current strains/inoculant strategies
  - Can more effective strains or strain variants be found for current applications?
  - Will genetic engineering of microbes and plants be useful for enhancing biocontrol?
  - How can formulations be used to enhance activities of known biocontrol agents?
- Discovering novel strains and mechanisms of action
  - Can previously uncharacterized microbes act as biological control agents?
  - What other genes and gene products are involved in pathogen suppression?
  - Which novel strain combinations work more effectively than individual agents?
  - Which signal molecules of plant and microbial origin regulate the expression of bioconitrol traits by different agents?
- Practical integration into agricultural systems
  - Which production systems can most benefit from biocontrol for disease management?
  - Which biocontrol strategies best fit with other IPM system components?
  - Can effective biocontrol-cultivar combinations be developed by plant breeders?

Over the past fifty years, academic research has led to the development of a small but vital commercial sector that produces a number of biocontrol products. The current status of commercialization of biological control products has been reviewed recently. As in most industries, funding in the private sector research and development goes through cycles, but seems likely to increase in the years ahead as regulatory and price pressures for agrochemical inputs increase.

Most of the commercial production of biological control agents is handled by relatively small companies, such as Agraquest, BioWorks, Novozymes, Prophyta, Kemira Agro. Occasionally, such companies are absorbed by or act as subsidiaries of multi-billion dollar agrochemical companies, such as Bayer, Monsanto, Syngenta, and Sumitomo.

Total revenues of products used for biocontrol of plant diseases represented just a small fraction of the total pesticide market during the first few years of the 21st century with total sales on the order of $10 to 20 million dollars annually. However, significant expansion is expected over the next 10 years due to increasing petroleum prices, the expanded demand for organic food, and increased demand for "safer" pesticides in agriculture, forestry, and urban landscapes.

Growers are interested in reducing dependence on chemical inputs, so biological controls (defined in the narrow sense) can be expected to play an important role in Integrated Pest Management (IPM) systems. A model describing the several steps required for a successful IPM has been developed. In this model, good cultural practices, including appropriate site selection, crop rotations, tillage, fertility and water management, provide the foundation for successful pest management by providing a fertile growing environment for the crop.

The use of pest- and disease-resistant cultivars, developed through conventional breeding or genetic engineering, provides the next line of defence. However, such measures are not always sufficient to be productive or economically sustainable. In such cases, the next step would be to deploy biorational controls of insect pests and diseases.

These include BCAs, introduced as inoculants or amendments, as well as active ingredients directly derived from natural origins and having a low impact on the environment and non-target organisms. If these foundational options are not sufficient to ensure plant health and/or economically sustainable production, then less specific and more harmful synthetic chemical toxins can be used to ensure productivity and profitability.

With the growing interest in reducing chemical inputs, companies involved in the manufacturing and marketing of BCAs should experience continued growth. However, stringent quality control measures must be adopted so that farmers get quality products. New, more effective and stable formulations also will need to be developed.

Most pathogens will be susceptible to one or more biocontrol strategies, but practical implementation on a commercial scale has been constrained by a number of factors. Cost, convenience, efficacy, and reliability of biological controls are important considerations, but only in relation to the alternative disease control strategies.

Cultural practices (e.g. good sanitation, soil preparation, and water management) and host resistance can go a long way towards controlling many diseases, so biocontrol should be applied only when such agronomic practices are insufficient for effective disease control. As long as petroleum is cheap and abundant, the cost and convenience of chemical pesticides will be difficult to surpass.

However, if the infection court or target pathogen can be effectively colonized using inoculation, the ability of the living organism to reproduce could greatly reduce application costs. In general, though, regulatory and cultural concerns about the health and safety of specific classes of pesticides are the primary economic drivers promoting the adoption of biological control strategies in urban and rural landscapes.

Self-perpetuating biological controls (e.g. hypovirulence of the chestnut blight pathogen) are also needed for control of diseases in forested and rangeland ecosystems where high application rates over larger land areas are not economically-feasible.

In terms of efficacy and reliability, the greatest successes in biological control have been achieved in situations where environmental conditions are most controlled or predictable and where biocontrol agents can preemptively colonize the infection court.

Monocyclic, soilborne and postharvest diseases have been controlled effectively by biological control agents that act as bioprotectants (i.e. preventing infections). Specific applications for high value crops targeting specific diseases (e.g. fireblight, downy mildew, and several nematode diseases) have also been adopted.

As research unravels the various conditions needed for successful biocontrol of different diseases, the adoption of BCAs in IPM systems is bound to increase in the years ahead.

Chapter 10

Fungicides and Germicides for Plant Diseases

Plant diseases are caused by parasitic fungi or by bacteria, or other vegetable parasites; or by forms of physiological disturbance. Each disease calls for special treatment. Most plant diseases must be prevented, not cured. It should be understood that spraying is only one of the control measures effective against plant diseases. Many diseases are not affected by spraying, though perhaps more are susceptible to this treatment than to any other.

A satisfactory fungicide must be one that does not injure the plants and at the same time is effective against the parasite. For spraying, additional requirements are imposed; it should not dissolve readily in rain water; it should adhere to foliage and fruit; in some cases it should be colorless in order not to make ornamentals more unsightly than when diseased.

The fungicide which has been used most for general purposes is bordeaux mixture. Lately some other preparations, particularly lime-sulfur combinations, have come into use, and in many cases are supplanting bordeaux. There are in addition a large number of other substances which have fungicidal value and are in more limited use for specific cases.

GENERAL PRACTICES

DESTROYING AFFECTED PARTS

It is important that all affected parts should be removed

and burned, if possible. In the fall all leaves and fruit that have been attacked by fungi should be raked up and burned. Diseased branches should be severed at some distance below the lowest visible point of attack. Fungous diseases often spread rapidly, and prompt action is usually necessary. Practice clean and tidy culture.

ROTATION OF CROPS

One of the most effective and practical means of heading off fungous diseases. Especially applicable to diseases of roots or root-crops, but also to many other diseases of annual plants.

STERILIZING BY STEAM

An effective fungicidal practice for several soil-inhabiting organisms which attack roots and stems. This includes nematode worms. It is especially applicable in the greenhouse, where it may be applied through sub-irrigation tile or through specially laid perforated steam pipes in the bottom of the bed. Cover the beds with blankets, introduce steam under pressure of 40 to 80 pounds for two hours. Insert thermometers at various places to see that the soil is being uniformly heated. A large galvanized iron tight box may be constructed with finely perforated trays 4 to 6 inches in depth. Soil placed in these trays and steamed for two hours as above will be freed from parasitic organisms. In this case the frames should be sprayed with a solution of formalin, 1 pint in 10 gallons of water.

Steam sterilization of soil may be used on intensively cultivated areas or extensive seed-beds. A portable boiler is necessary. The beds are sterilized after they have been prepared for seed, and just before the seed is sown. A galvanized pan 10 by 6 feet and 6 inches deep is inverted, and the edges are pushed down into the soil one or two inches.

The pan is connected with the steam boiler by means of a steam hose and live steam is run into the pan from about forty minutes under a pressure of 100 pounds and up. The higher the pressure the more thoroughly the soil will be sterilized. The cost of sterilizing is approximately three-fourths of a cent the square foot. It should be noted that soil sterilization has an invigorating effect on the plants, and it will be necessary to run greenhouses at a lower temperature (5°–10°) both night and day. Field sterilization also kills weed seeds, and with the reduction of the cost of weeding makes the process practicable.

CHEMICAL SUBSTANCES

BORDEAUX MIXTURE

A bluish-green copper compound that settles out when freshly slaked lime and a solution of copper sulfate are mixed. Many formulas have been recommended and used. The 5-5-50 formula may be regarded as standard.

In such a formula the first figure refers to the number of pounds of copper sulfate, the second to the stone or hydrated lime, and the third to the number of gallons of water. Bordeaux must often be used as weak as 2–2–50, on account of injury to some plants.

To make 50 gallons of bordeaux mixture, proceed as follows:

- Pulverize 5 pounds of copper sulfate (blue vitriol), place in a glass, wooden, or brass vessel, and add two or three gallons of hot water. In another vessel slake 5 pounds of quicklime in a small amount of water. When the copper sulfate is all dissolved, pour into a barrel and add water to make 40 or 45 gallons. Now strain the lime into this, using a sieve 50 meshes to the inch or a piece of cheese-cloth supported by ordinary screening. Stir thoroughly, and add water to the 50-gallon mark. The flocculent substance which settles is the effective fungicide. Always stir vigorously before filling the sprayer. Never add the strong lime to strong vitriol. Always add a large amount of water to one or the other first. Blue vitriol used alone would not only wash off quickly in a rain, but cause a severe burning of fruit and foliage. Lime is added to neutralize this burning effect of the copper. If the lime were absolutely pure only slightly more than one pound would be required to neutralize this burning effect. For many purposes an excess of lime is not objectionable and may be desirable. For nearly ripe fruit and ornamentals an excess of lime augments spotting. In such cases the least amount of lime possible should be used. Determine this by applying the cyanide test.

- Secure from the druggist 10 cents' worth of potassium ferrocyanide (yellow prussiate of potash) and dissolve it in water in an 8-ounce bottle. Cut a V-shaped slit in one side of the cork, so that a few drops of the liquid can be obtained. Now proceed as before. Add lime with constant stirring until a drop of the ferrocyanide ceases to give a reddish-brown colour.
- When bordeaux mixture is desired in large quantities, stock solutions should be made. Place 100 pounds of copper sulfate in a bag of coffee-sacking, and suspend in the top of a 50-gallon barrel, and add water to the 50-gallon mark. In twelve to fifteen hours the vitriol will be dissolved and each gallon of solution will contain 2 pounds of copper sulfate.

  Slake a barrel of lime, and store in a tight barrel, keeping it covered with water. Lime so treated will keep all summer. It is really hydrated lime. This is often dried, pulverized, and offered on the market in paper bags of 40 pounds, each, under such names as ground lime, prepared lime, hydrated lime, etc. If the paper is not broken, the lime does not air-slake for a long time. One and one third pounds of hydrated lime equals in value one pound of quicklime. Air slaked lime cannot be used in preparing bordeaux mixture.

Arsenical poisons can be combined with bordeaux mixture.

AMMONIACAL COPPER CARBONATE

For use on nearly mature fruit and on ornamentals. Does

not discolor. Weigh out 3 ounces of copper carbonate, and make a thick paste with water in a wooden pail. Measure 5 pints of strong ammonia and dilute with three or four parts of water. Add ammonia to the paste, and stir. This makes a deep blue solution. Add water to make 50 gallons.

COPPER CARBONATE

For use in the above formula, it may be obtained as a green powder, or may be prepared as follows: Dissolve 12 pounds of copper sulfate in 12 gallons of water in a barrel. Dissolve 15 pounds of sal soda in 15 gallons of water (preferably hot). Allow the solution to cool; then add the sal soda solution to the copper sulfate solution, pouring slowly in order to prevent the mixture from working up and running over. A fine precipitate is formed which will settle to the bottom if allowed to stand over night. Siphon off the clear liquid. Wash the precipitate by adding clear water, stirring, and allowing to settle. Siphon off the clear water, strain the precipitate through muslin, and allow it to dry. This is copper carbonate. The above amounts will make about 6 pounds.

CORROSIVE SUBLIMATE (MERCURIC BICHLORIDE)

Used for disinfecting pruned stubs and cleaned-out cankers, at the rate of one part in 1000 parts of water. Can be secured from the druggist in tablet form in vials of 25 each, and costing 25 cents: One tablet makes a pint of solution. Make and store solution in glass and label poison.

FORMALIN

A pungent, clear liquid, very irritating to eyes and nose. Obtained at any drug store at about 40 cents per pint. Used for potato-scab, oat smut, bunt in wheat, soil disinfection, etc.

LIME

Offered for sale in the following forms:

- Ground rock or ground limestone; air-slaked lime is of the same composition, i.e. a carbonate of calcium.

- Lump, barrel, stone, or quick lime; this is burned limestone, and should preferably test 90 per cent oxide of calcium.
- Prepared, ground, or hydrated lime; this is water or steam-slaked quicklime, dried and pulverized.

Used as an applicant to the soil to correct acidity, for club-root of cabbage, etc., and for preparing spray mixtures.

LIME-SULFUR

In the many possible combinations, lime-sulfur is coming to be equally as important as bordeaux mixture, in the control of many plant diseases.

- A mixture of equal parts of dry lime and powdered sulfur is often dusted on plants for surface mildews.
- A paste of equal parts of lime, sulfur, and water. This is painted on the heating pipes in the greenhouse, and is valuable for keeping off surface mildews.
  - Home-boiled dilute lime-sulfur. This solution has been widely used in the past as a dormant spray, particularly for San Jose scale and peach leaf-curl. It is likely to be supplanted.
  - Home-boiled concentrated lime-sulfur.— When a great deal of spraying is to be done, a concentrated lime-sulfur solution may be boiled at home and stored in barrels to be used as needed.

Test with a Beaume hydrometer, which has a scale reading from 25° to 35°. Dilutions are reckoned from a standard solution testing 32°. If the solution tests only 28°, it is not as strong as standard, and cannot be diluted as much as a solution testing 32°. Decimals are given in ail cases, but for practical purposes the nearest even gallon or half gallon can be used, unless appliances for more accurate measurement are at hand. It is understood in making all dilutions that water is added to one gallon of the concentrate to make the stated amount. Do not measure out the stated amount of water and add the concentrated solution to it.

	1-10	1-15	1-20	1-25	1-30	1-40	1-50	1-60	1-75	1-100
25°	7.4	11	14.7	18.4	22.1	29.5	36.8	44.2	55	73
26°	7.7	11.6	15.4	19.3	23.2	30.9	38.6	46.3	58	77.2
27°	8.1	12.1	16.1	20.2	24.3	32.4	40.5	48.5	60.6	80.7
28°	8.4	12.7	16.9	21.1	25.4	33.8	42.3	50.7	63.5	84.5
29°	8.8	13.2	17.6	22.1	26.5	35.3	44.2	53	66.3	88.2
30°	9.2	13.9	18.4	23	27.6	36.9	46.1	55.3	69	92
31°	9.6	14.4	19.3	24	28.8	38.4	48	58	72	96
32°	10	15	20	25	30	40	50	60	75	100
33°	10.4	15.6	20.8	26	31.2	41.5	52	62.4	78	104
34°	10.8	16.2	21.6	26.8	32.4	43.2	54	64.7	80.8	108
35°	11.2	16.8	22.4	28	33.4	44.9	56	67.4	84.2	112

— Commercial concentrated lime-sulfur. - As manufactured and placed on the market is a clear amber liquid, and should test 32° to 35° Beaume It costs about 20 cents per gallon retail, and comes ready to pour into the spray tank. For apple and pear diseases. Arsenate of lead can be used with this solution, and increases its fungicidal value.

— Self-boiled lime-sulfur. This is a mechanical mixture of the two substances, and is really not boiled, the heat being supplied by the slaking lime. In a small barrel or keg place 8 pounds of good quicklime.

Add water from time to time in just sufficient amounts to prevent burning. As soon as the lime begins to slake well, add slowly (preferably through a sieve) 8 pounds of sulfur flour. Stir constantly, and add water as needed. As soon as all bubbling has ceased, check further action by adding a quantity of cold water, or pour into a barrel or tank and make up to 50 gallons. Keep well agitated. Very effective against peach scab and brown rot. Several other formulas have been used: 10–10–50 and 5–5–50. Arsenate of lead can be used with this mixture.

By using boiling water and allowing the hot mixture to

stand for half an hour, a stronger spray mixture of the above can be secured. It cannot be used safely on peaches, but has been used successfully on grapes for surface mildew. The addition of sulfate of iron or sulfate of copper, one or two pounds to 50 gallons, has been used for apple rust.

POTASSIUM SULFID (LIVER OF SULFUR)

Simple solution 3 ounces in 10 gallons of water. For mildew in greenhouses, on rose-bushes and other ornamentals.

RESIN-SAL-SODA STICKER

Resin, 2 pounds; sal soda (crystals), 1 pound, water, 1 gallon. Boil until of a clear brown colour, i.e. from one to one and a half hours. Cook in an iron kettle in the open. Add this amount to 50 gallons of bordeaux. Useful for onions, cabbage, and other plants to which spray does not adhere well.

SULFATE OF COPPER (BLUE VITRIOL)

Dissolve 1 pound of pure sulfate of copper in 25 gallons of water. A specific for peach leaf-curl. Apply once before buds swell in the spring. Cover every bud. For use in preparing bordeaux mixture. Costs from 5 to 7 cents per pound, in quantity.

SULFATE OF IRON (COPPERAS)

A greenish granular crystalline substance. Dissolve 100 pounds in 50 gallons of water. For mustard in oats, wheat, etc., apply at the rate of 50 gallons per acre. Also for anthracnose of grapes as a dormant spray.

SULFUR

Should be 99 per cent pure. Valuable for surface mildews. Dust on dry or in the greenhouse used in fumes. Evaporate it over a steady heat, as an oil stove, until the house is filled with vapour. Do not heat to the burning point, as burning sulfur destroys most plants. To prevent burning, place the sulfur and pan in a larger pan of sand and set the whole upon the oil stove.

Chapter 11

Insects Affected Plants and Their Control

APPLE

Apple-bucculatrix (Bucculatrix pomifoliella): A minute yellow or green larva feeding upon the upper surface of the leaves, causing the lower surface to turn brown. The cocoons are white and slender, and are laid side by side upon the under side of twigs, where they are conspicuous in winter.

Treatment: Lime-sulfur while tree is dormant. Arsenicals for the larvae in summer.

Apple-curculio (Anthonomus quadrigibbus): A soft, white grub, about half an inch long, living in the fruit.

Treatment: Clean cultivation. Rake the small apples that drop early out into the sun where they will dry up.

Apple Flea-beetle (Graptodera foliacea): Brassy, green beetle, one-fifth inch or less long, feeding upon leaves.

Treatment: Arsenicals. Lime-sulfur or bordeaux mixture as a repellent.

Apple-Maggot or Railroad-worm (Rhagoletis pomonella): Maggot; infests harvest and fall apples mostly, occasionally attacks winter fruit. It tunnels apples through and through, causing the fruit to fall to the earth.

Treatment: Pick up all windfalls every two or three days, and either feed them out or bury them deeply, thus killing the maggots. Pasture to hogs.

Bud-moth (Tmetocera ocellana): The small brown

caterpillars with black heads devour the tender leaves and flowers of the opening buds in early spring.

Treatment: Make two applications of either 1 pound paris green or 4 pounds arsenate of lead in 100 gallons of water; the first when the leaf-tips appear, and the second just before the blossoms open. If necessary, spray again after the blossoms fall. In cases where lime-sulfur is used just before the buds open for scale or blister mite, arsenate of lead, 4 pounds to 100 gallons, may be added and will help to control the bud-moth.

Case-bearers: **The pistol-case-bearer (Coleophora malivorella) and the cigar-case-bearer (C. fletcherella):** The small caterpillars live in pistol or cigar-shaped cases, about a quarter of an inch long, that they carry around with them. They appear in spring on the opening buds at the same time as the bud-moth, and may be controlled by the same means.

CANKER-WORM

Spring and fall (Paleacrita vernata and Alsophila pometaria): Larva; a " measuring worm," an inch long, dark, and variously striped, feeding upon the leaves.

Preventive: Band the trees with tanglefoot to prevent the wingless females from climbing.

Treatment: Arsenicals, thoroughly applied in spray, are very effective.

Codlin-moth (Carpocapsa pomonella): This is the pinkish caterpillar which causes a large proportion of wormy apples. The eggs are laid by a small moth on the leaves and the skin of the fruit. Most of the caterpillars enter the apple at the blossom end.

When the petals fall, the calyx is open, and this is the time to spray. The calyx soon closes, and keeps the poison inside ready for the young caterpillars' first meal. After the calyx has closed, it is too late to spray effectively. The caterpillars become full grown in July and August, leave the fruit, crawl down on the trunk, and there most of them spin cocoons under the loose bark. In most parts of the country there are two broods annually.

Treatment: When the majority of the petals have fallen, spray with 1 pound paris green or 4 pounds arsenate of lead in 100 gallons of water, using a stiff spray to force it into the blossom end of the apple. Repeat the application three weeks later. For use of the poison with bordeaux or lime-sulfur.

Fall Web-worm (Hyphantria cunea): Hairy larva, about an inch long, varying from gray to pale yellow or bluish black, feeding upon the leaves of many trees, in tents or webs.

Treatment: Destroy by burning the webs, or removing them and crushing the larvae. Spray with arsenicals.

Leaf Blister Mite (Eriophyes pyri): The presence of this minute mite is indicated by small irregular brownish blisters on the leaves.

Treatment: Spray in late fall or early spring with lime-sulfur, or miscible oil.

Flat-headed Borer (Chrysobothris femorata): Larva about an inch long, flesh-colored, the second segment (" head ") greatly enlarged; boring under the bark and sometimes into the wood. They are readily located in late summer or fall by the dead and sunken patches of bark.

Preventive: Soap and carbolic acid washes applied from May to July. Keep trees vigorous.

Treatment: Dig out the borers in early summer and fall. Encourage woodpeckers.

PEAR

Blight (Boxillus amylovorus): A very serious bacterial disease. Bacteria winter just at the edge of the dead wood in trees blighted the previous year. With the advent of warm spring days they ooze through the bark in sticky drops and are carried by bees and flies to blossoms. The blossoms blight, and the spur may also blight. Plant-lice carry bacteria from blighted blossoms to spurs and shoots.

If a spur becomes blighted, the bacteria may spread in the bark of the limb, causing a depression or canker. This may girdle the limb and cause its death. The leaves turn black and stick tenaciously, even through the winter. Succulent water sprouts are very apt to blight and cause large cankers.

Generally distributed in North America, and known only in America. Attacks apple, quince, mountain ash, hawthorn; the Spitzenburgh is specially liable to attack.

Treatment: Clean up hedgerows of hawthorn, old blighted pear trees and apple trees. In early spring cut out the blight of the previous year and disinfect the stubs with corrosive sublimate, 1-1000. Clean out cankers with a sharp knife, and disinfect. Paint over with lead paint. At blossoming time make a systematic daily inspection for blossom blight, and break it out.

Watch for blight in the shoots. When it appears get a long-handled pruning-hook, fasten a sponge near the knife, and saturate it with corrosive sublimate solution, 1-1000. Clip out the blighted twigs, cutting five or six inches below the blight, and sop the pruned stub with the sponge. During a blight epidemic, drop all other work. The work must be done systematically and persistently, or not at all. One week's work may save the pear crop and the pear trees.

Leaf-blight and Cracking of fruit (Fabrea maculata): Attacks nursery stock of pears and quinces, beginning as small circular brown spots on the leaves. These spread, and if numerous cause the leaf to fall. The same disease produces a black spot or pit on the fruit.

Treatment: In nurseries spray with bordeaux mixture, 4-4-50. In the orchard spray as for pear scab, with perhaps one additional application.

Leaf-spot (Mycosphaerella sentina): Small lecticular spots with white centers on leaves. Spots become so numerous as to cause defoliation. The fungus is known only on leaves, and it winters on them.

Treatment: Burn fallen leaves. Owing to the nearness of the fungus (on the twigs) and the slowness with which the pear-leaf unfolds, two applications of spray before the blossoms open are sometimes necessary, and one immediately after they fall. Use lime-sulfur, 1-50, or bordeaux, 3-3-50.

Scab (Venturia pyrina): Greenish brown or black spots on leaves and fruit, arresting growth and often causing fruit to crack. Severe on Flemish Beauty. Often attacks pedicles of fruits and causes them to drop, and may even cause defoliation. Is different from apple scab, but behaves much like it. Differs especially in the fact that the fungus winters on the twigs as well as on fallen leaves.

Treatment: Owing to the nearness of the fungus (on the twigs) and the slowness with which the pear-leaf unfolds, two applications of spray before the blossoms open are sometimes necessary, and one immediately after they fall. Use lime-sulfur, 1-50, or bordeaux, 3-3-50.

Rust (Gymnosporangium globosum): Having the same habits and appearance as apple rust.

Treatment: Owing to the nearness of the fungus (on the twigs) and the slowness with which the pear-leaf unfolds, two applications of spray before the blossoms open are sometimes necessary, and one immediately after they fall. Use lime-sulfur, or bordeaux.

Plum-curculio (Conotrachelus nenuphar): Beetle; deforms the fruit by its characteristic feeding and egg-laying punctures. The grubs develop in the fruit and cause it to fall.

Treatment: Spraying with arsenate of lead, as for codlin-moth, whenever it can be applied with a fungicide so as not to increase expense, will help to control the trouble. Thorough. superficial tillage of the surface soil during July and August will kill many of the pupae, and is recommended.

Round-headed Borer (Saperda Candida): A yellowish white larva, about one inch long when mature. It is said to remain in the larval state three years.

Preventive: Keep the beetles from laying eggs by spraying the trunks several times during the spring and summer with kerosene emulsion or by coating them with an alkaline wash made from soap, caustic potash, and carbolic acid. Tarred paper tree-protectors well tied at the top, or wire mosquito netting protectors closed at the top and encircling the trunk so loosely that the beetles cannot reach the bark, are effective in preventing egg-laying. Practice clean cultivation, and do not let water sprouts or other rank vegetation encircle the base of the tree.

Remedial: Dig out the borers whenever they can be

located by discolored bark or by the sawdust thrown out of the burrow.

Leaf-crumpler (Mineola indigenella): Reddish brown caterpillars that live in slender, horn-shaped cases and feed on the tender leaves. They hibernate as partly grown larvae and attack the opening buds the following spring. They usually live in a nest of several leaves fastened together with silk.

Treatment: Gather the nests and burn them. Arsenicals when the buds open.

Oyster-shell Scale (Lepidosaphes ulmi): This is an elongate scale (sometimes called bark-louse), one-eighth inch in length, resembling an oyster-shell in shape and often incrusting the bark. It hibernates as minute white eggs under the old scales. The eggs hatch during the latter part of May or in June, the date depending on the season. After they hatch, the young may be seen as tiny whitish Apple Insects lice crawling about on the bark. When these young appear, spray with kerosene emulsion, diluted with 6 parts of water, or whale-oil, or any good soap, 1 pound in 4 or 5 gallons of water. Where trees are regularly sprayed with lime-sulfur as for the San Jose scale or blister mite, the oyster-shell scale is usually controlled.

Scurfy Scale (Chionaspis furfurus): This whitish, pear-shaped scale, about one-eighth inch in length, often incrusts the bark, giving it a scurfy appearance. It hibernates as purplish eggs under the old scales.

Treatment: Spray as recommended for Oyster-shell Scale.

Tent-caterpillars (Malacosoma americana and M. disstria). -Larva, nearly two inches long, spotted and striped with yellow, white, and black; feeding upon the leaves. They congregate in tents or in clusters on the bark at night and in cool weather, and forage out upon the branches during the day.

Treatment: Arsenicals, as for Codlin-moth. Burn out nests with torch, or cut them out and crush the larvae. Pick off egg masses from twigs during winter and spring.

Tussock-moth (Hemerocampa leucostigma): A handsome, redheaded, yellow and black tufted caterpillar, about an inch long, which devours the leaves and sometimes eats into the fruit.

Remedial: Collect the frothy egg-masses in fall and winter and band the trees to prevent a reinfestation by migrating caterpillars. Spray with arsenicals as for codlin-moth, taking care to cover the under side of the leaves.

Twig-borer (Schistoceros hamatus): Beetle, three-eighths inch long, cylindrical and dark brown, boring into twigs of apple, pear, and other trees. The beetle enters just above a bud.

Treatment: Burn the twigs. The early stages are passed in dying wood such as prunings, diseased canes, and in upturned roots. Burn such rubbish, and thus destroy their breeding-places. This is also a grape pest.

Twig-Pruner (Elaphidion villosum): Yellowish white larvae, about a half inch long, boring into young twigs, causing them to die and break off.

Treatment: Burn the twigs.

Woolly Aphis {Schizoneura lanigera): Small reddish-brown plant-lice covered with a conspicuous mass of white, waxy fibers, found on the branches, sprouts, trunks, and roots.

Preventive: Do not set infested trees.

Treatment: For the form above ground drench the infested parts with 15 per cent kerosene emulsion; for the underground form remove the earth beneath the tree to a depth of 3 inches, and apply 10 per cent kerosene emulsion liberally, and replace the earth. In the case of nursery stock the emulsion may be applied in a shallow furrow close to the row.

Brown Apricot-scale (Eulecanium armeniacum): A soft brown scale infesting the under side of the smaller branches.

Treatment: Spray with resin and fish-oil compound, taking care to hit the underside of the twigs. In California the application should be made in January and February.

ASPARAGUS

Common Asparagus-beetle (Crioceris asparagi): Beetle, less than one-fourth inch in length, yellow, red, and shining black, with conspicuous ornamentation, feeding upon the tender shoots. Larva feeds upon the leaves and tender bark.

Treatment: Freshly slaked lime dusted on before the dew

has disappeared in the morning. Poultry. Cut down all plants in early spring to force the beetles to deposit their eggs upon the new shoots, which are then cut every few days before the eggs hatch; or leave a row or so around the field as a lure for the beetles where they may be killed with arsenicals. The Twelve-spotted Asparagus-beetle. Similar to the last, but with twelve spots on the wing-covers.

Treatment: Similar to that used above, except that the grubs cannot be destroyed by lime, since they live within the berry.

Asparagus Miner (Agromyza simplex): A maggot mining under the skin near the base of the plant.

Treatment: Leave a few volunteer plants as a trap in which the fly will deposit her eggs. Pull and burn these plants in late June and early July.

ASTER

Aster-worm (Papaipema nitela): A small larva boring in the stem of garden asters about the time they begin to flower, causing the heads to droop.

All infested stocks should be burned. Destroy by burning all rank weeds, such as ragweed and cocklebur, before September.

BEAN

Bean-weevil or Bean-bug (Bruchus obtectus): Closely resembles the pea-weevil, which see for description and remedies. Holding over the seed will be of no value with this insect.

Seed-corn Maggot. (Pegomya fusciceps): A maggot attacking germinating seeds and roots of young plants.

Treatment: Avoid stable manure; practice crop rotation. In the garden use sand moistened with kerosene around the plants to keep the flies from laying the eggs.

BIRCH

Bronze Birch-borer (Agrilus anxius): A slender, creamy white grub, three-fourths inch in length when full grown, that burrows under the bark of the white birch, ultimately killing the tree. The eggs are laid during May and June by a slender, olive-bronze beetle about one-half inch in length.

Treatment: After a tree has become thoroughly infested, nothing can be done to save it. As the first indication of the presence of the borer is usually a dying of the topmost branches, such trees should be carefully examined, and if infested should be cut down and burned before May 1, to prevent a spread of the trouble to other trees.

CABBAGE

Cabbage-worm or Cabbage-butterfly (Pontia rapae): The green caterpillars hatch from eggs laid by the common white butterfly. There are several broods every season.

Treatment: If plants are not heading, spray with kerosene emulsion or with paris green to which the sticker has been added. If heading, apply hellebore.

Common Cabbage-looper (Autographa brassicae): A pale green caterpillar, striped with lighter lines. Feeds on the leaves.

Treatment: Arsenicals applied to lower surface of leaves.

Cabbage Aphis (Aphis brassicae): These small, mealy plant-lice are especially troublesome during cool, dry seasons, when their natural enemies are less active.

Treatment: Before the plants begin to head, spray with kerosene emulsion diluted with 6 parts of water or whale-oil soap, 1 pound in 6 gallons of water, or use one of the concentrated tobacco extracts. Destroy all cabbage stalks and other cruciferous plants in the fall. Dip infested plants in soap solution before planting.

Harlequin Cabbage-bug (Murgantia histrionica): Bug about a half-inch long, gaudily colored with orange dots and stripes over a blue-black ground, feeding upon cabbage; two to six broods.

Treatment: Hand-picking. Place blocks about the patch, and the bugs will collect under them. In the fall make small piles of the rubbish in the patch, and burn them at the approach of winter. Practice clean culture. Destroy all cabbage stalks and other cruciferous plants in fall. Early in the spring plant a trap crop of mustard, radish, rape, or kale. When the overwintering bugs congregate on these plants, destroy them with pure kerosene or by hand.

Maggot (Pegomya brassicae): A minute white maggot, the larva of a small fly, eating into the crown and roots of young cabbage, cauliflower, radish, and turnip plants.

Treatment: Carbolic acid emulsion diluted with 30 parts of water applied the day following the transplanting of the cabbage plants, and repeated once a week for several applications. Remove a little earth from about the plants, and spray on the emulsion forcibly. It has also been found practicable to protect the plants by the use of tightly fitting cards cut from tarred paper.

In seed beds protect the plants by surrounding the bed with boards one foot wide placed on edge, across which a tight cover of cheese-cloth is stretched.

CARROT

Carrot-beetle (Ligyrus gibbosus): A reddish brown beetle one-half inch or more long, which attacks the young plants. The larva lives in the ground, where it feeds on humus.

Preventive: Crop rotation and other remedies for white grub.

CAULIFLOWER

Carrot Rust-fly (Psila rosae): Minute whitish yellow maggots infesting the roots and stunting the plants.

Preventive: Late sowing and rotation of crops. Celery or carrots should not follow each other.

Celery Caterpillar (Papilio polyxenes): A large green caterpillar, ringed with black and spotted with yellow, which feeds on the leaves.

Treatment: Hand-picking as soon as observed.

Celery Leaf-tyer (Phlycaenia ferrugalis): A greenish caterpillar, feeding on the under side of the leaves.

Treatment: Spray with arsenicals while the larvae are still young.

Little Negro-bug (Corimelaena pulicaria): Glossy black

bugs one-eighth inch in length, which collect in clusters in the axils of the leaflets and cause the plants to wilt.

Treatment: Kerosene emulsion or tobacco extract.

CHERRY

Slug (Eriocampoides limacina): Larva; one-half inch long, blackish and slimy, feeding upon the leaves; two broods.

Treatment: Arsenicals, hellebore, tobacco extract.

Aphis (Myzus cerasi): Blackish plant lice infesting the leaves and tips of new growth.

Treatment: Spray as soon as the first lice appear with whale-oil soap or tobacco extract.

CHEST NUT

Weevil (Balaninus proboscideus and B. rectus): A grub working in chestnuts, making them wormy. The weevil is a curculio-like insect.

Preventives: Destroy wild trees where the insects breed. Plant the most immune varieties. Gather and destroy the infested nuts immediately after they fall.

Chrysanthemum Leaf-miner (Oscinis sp.): Works on the leaves of the chrysanthemum.

Remedy: Hand-picking.

CLOVER

Flower-midge (Dasyneura leguminicola): An orange-red maggot infesting the flower-buds, where they consume the contents of the ovary.

Preventives: Cut the first crop for hay as early as possible, thus destroying the undeveloped larvae of the first brood.

Seed-chalcis (Bruchophagus funebris): A white grub found inside the seed.

Preventive: Same as for Flower-midge, above. Destroy all volunteer clover plants.

Seed-caterpillar (Enarmonia interstictana): A small whitish or orange caterpillar infesting the heads.

Preventive: Early cutting of first crop, as for Flower-midge.

Root-borer (Hylastinus obscurus): Small white grub burrowing in the roots.

Preventive: Plow under badly infested fields as soon as possible after cutting.

Hay-worm (Hypsopygia costalis): A brownish caterpillar three-fourths inch long, infesting stacked or stored clover.

Preventive: Remove old clover hay before putting in the new. Place stacks on log or rail foundation, and salt the lower layers.

CORN

Corn-root Aphis(Aphis maidiradicis): A bluish green aphis infesting the roots.

Preventives: A short rotation period in corn, especially in dry years. Deep and thorough and repeated stirring of old corn ground in fall and spring as a preparation for corn-planting. Maintenance and increase of the fertility of the soil.

White Grubs (Lachnosterna spp.): The large white curved larvae of the common June beetle.

Preventives: Rotation of crops; do not let corn follow sod, but let a crop of clover or clover and oats intervene. To help clear sod land of grubs, pasture to hogs any time between April and October. To prevent laying of eggs in corn-field, keep the ground free from weeds during May and June. Thorough cultivation and heavy fertilization.

Northern Corn Root-worm (Diabrotica longicornis): A whitish grub two-fifths inch long, which burrows in the roots.

Preventive: Crop rotation; corn should not follow corn

Wire-worms (Elateridae): Hard, yellowish, or reddish, cylindrical larvae feeding on the roots.

Preventives: Crop rotation; let clover intervene between sod and corn, planting the corn late the second or third year. Early fall plowing.

Cut-worms (Agrotis, Hadena, etc.): Soft-bodied caterpillars eating and cutting off the young plants.

Preventives: Early fall plowing of grass lands intended for corn; pasturing by pigs of grass or clover land intended for corn; distributing a line of poisoned bran by means of a seed-drill. To prevent the caterpillars entering from a neighboring grass field, destroy them with a line of poisoned vegetable bait.

Sod Web-worms (Crambus spp.): Gray or brownish caterpillars about one-half inch long, living in a silk-lined burrow in the soil at base of the plant. They thrive in grass land.

Preventive: Early fall plowing of grass land intended for corn, or else plow as late as possible the next spring.

Army-worm. (Leucania unipuncta): A cut-worm-like caterpillar, which normally feed on grass. When this food supply is exhausted, they migrate in numbers to other fields and attack corn, wheat, etc.

Preventive: To stop the advance of the " army," plow deep furrows so the dirt is thrown towards the colony; in the bottom of the furrow dig post holes into which the caterpillars will fall and where they may be killed with kerosene.

Chinch-bug (Blissus leucopterus): A red or white and black sucking bug, three-twentieths of an inch long. Attacks wheat and corn in great numbers.

Preventives: Clean farming to destroy suitable hibernating shelter. Stop the migration of the bugs from the wheat-fields into corn by maintaining along the field a dust strip ten feet wide in which a furrow and post-hole barrier has been constructed. This may be supplemented by a coal-tar barrier.

Grasshoppers (Acrididae): Poison them with the Criddle mixture.

Corn Ear-worm (Heliothis armiger): A green or brownish striped caterpillar feeding on the corn beneath the husk. Three to six generations yearly.

Preventives: Plant as early as possible, and still avoid a "set back " to the crop.

COTTON

Bollworm (Heliothis obsoleta): This insect is also known as the corn earworm and tomato fruit-worm. The caterpillars are over an inch in length, and vary in colour from greenish to dark brown.

Preventives: Produce an early crop of cotton by planting early varieties, heavy fertilizing, early and frequent cultivation. Practice fall plowing, to destroy as many hibernating pupae as possible. Use corn as a trap crop.

Plant it in strips across the field and time it so that the crop will be in silk and tassel about August 1. In areas infested by the boll weevil follow the recommendations given below.

Mexican Boll-weevil (Anthonomus grandis): A snout beetle about one-fourth inch in length, which lays its eggs in the squares and bolls, producing a grub which eats out the contents.

Treatment:

- Destroy the vast majority of weevils in the fall by uprooting and burning the plants. This is the all-important step. It results in the death of millions of weevils. It insures a crop for the following season.
- Destroy also many weevils that have survived the preceding operation and are found in the cotton-fields and along the hedgerows, fences, and buildings. This is done by clearing the places referred to thoroughly.
- As far as possible, locate the fields in situations where damage will be avoided. This cannot be done in all cases, but can frequently be done to good advantage.
- Prepare the land early and thoroughly in order to obtain an early crop. This means fall plowing and winter working of the land.
- Provide wide rows, and plenty of space between the rows and the plants in the drill, for the assistance of the natural enemies of the weevil, which do more against the pest than the farmer can do himself by any known means. Check-rowing, wherever practicable, is an excellent practice.
- Insure an early crop by early planting of early-maturing varieties, and by fertilizing where necessary.

- Continue the procuring of an early crop by early chopping to a stand and early and frequent cultivation. Do not lose the fruit the plants have set by cultivation too deep or too close to the rows.
- Where the labour is sufficient, pick the first appearing weevils and the first infested squares. Do not destroy the squares, but place them in screened cages.
  By this means the escape of the weevils will be prevented, while the parasites will be able to escape to continue their assistance on the side of the farmer.
- Use a crossbar of iron or wood, or some similar device, to cause the infested squares to fall early to the ground, so that they will be exposed to the important effects of heat and parasites.
- Do not poison for the leaf-worm unless its work begins at an abnormally early date in the summer.

CRANBERRY

Fruit-worm (Mineola vaccinii): Small caterpillar working in the fruits, eating out the insides.

Preventive: For bogs with abundant water, reflow for ten days immediately after picking. Let the foliage ripen, and then turn on water for winter. Draw off water early in April, and every third or fourth year hold it on until the middle of May. For dry bogs spray three times with arsenate of lead during July. Bury all screenings.

Cranberry-worm (Eudemis vacciniana): Small larva, green, black-headed, feeding upon the shoots and young leaves, drawing them together by silken threads; two broods.

Treatment: Flooding for two or three days when the worms come down to pupate.

ARSENICALS

Yellow-headed Cranberry-worm (Acleris minuta): Stout, yellowish-green, small caterpillar, with a yellow head, webbing up the leaves as it works.

Treatment: Hold the water late on the bog in spring to prevent egg-laying. Arsenicals from the middle of May till July 1.

Cranberry-girdler (Crambus hortuellus): Small caterpillars feeding on the stems just beneath the surface of the sand.

Preventive: Reflow just after picking, for a week or ten days, or reflow for a day or two about June 10.

False Army-worm (Calocampa nupera): Green to blackish caterpillars devouring the leaves and buds.

Treatment: Reflow for from twenty-four to thirty-six hours soon after the middle of May. It may be necessary to reflow a second time. Destroy all caterpillars washed ashore while the water is on.

In dry bogs, spray early in May with arsenate of lead.

CUCUMBER

Pickle-worm (Diaphania nitidalis): Larva, about an inch long, yellowish white, tinged with green, boring into cucumbers; two broods.

Preventives: Clean farming, fall plowing, and rotation of crops.

Remedies: Kill the caterpillars before they enter the fruit by spraying with arsenate of lead about the time the buds begin to form, and repeat in two weeks.

Spotted Cucumber-beetle (Diabrotica 12-punctata): Beetle, yellowish and black spotted, about one-fourth inch long, feeding upon the leaves and fruit. Sometimes attacks fruit-trees, and the larva may injure roots of corn.

Treatment: Same as for Striped Cucumber-beetle.

Striped Cucumber-beetle (Diabrotica vittata): Beetle, one-fourth inch long, yellow with black stripes, feeding on leaves. Larva one-eighth inch long and size of a pin, feeding on roots; two broods.

Preventive: Cheap boxes covered with thin muslin or screens of mosquito-netting, placed over young plants.

Remedies: Arsenicals in flour. Arsenate of lead. Ashes, lime, plaster, or fine road dust sprinkled on the plants every two or three days when they are wet. Air-slaked lime. Plaster and kerosene. Tobacco powder, applied liberally. Apply remedies when dew is on, and see that it strikes the under side of the leaves.

Currant. Borer (Sesia tipuliformis): A whitish larva, boring in the canes of currants, and sometimes of gooseberries. The larva remains in the cane over winter.

Treatment: In fall and early spring cut and burn all affected canes. These canes are distinguished before cutting by lack of vigour and by limberness.

Currant-worm, or Currant and Gooseberry Sawfly (Nematus ventricosus): Larva, about three-fourths inch long, yellow-green, feeding on leaves of red and white varieties; two to four broods.

Treatment: White hellebore, applied early. Arsenicals for the early brood. Treatment should begin while the larvae are on the lowermost leaves of the bushes.

Before the leaves are fully grown, the holes made by the worms may be seen. The second brood is best destroyed by killing the first brood.

Currant Measuring or Span-worm (Cymatophora ribearia): Larva somewhat over an inch long, with stripes and dotted with yellow or black, feeding upon the leaves.

Treatment: Hellebore, applied stronger than for currant-worm.

Green Leaf-hopper (Empoa albopicta): Small insect working upon the under surface of currant and gooseberry leaves. Also upon the apple.

Remedies: Pyrethrum. Kerosene emulsion. Tobacco-dust. Tobacco extracts.

EGG-PLANT

Elm Leaf-beetle (Galerucella luteola): A small beetle, imported from Europe, which causes great devastation in some of the eastern states by eating the green matter from elm leaves, causing the tree to appear as if scorched.

Remedy: Arsenate of lead (1/2 pounds to 25 gallons).

Elm Sawfly Leaf-miner (Kaliosysphinga ulmi): A greenish white larva feeding between the two layers of the leaf, causing large blotches; when abundant, the leaf dies and falls. They sometimes kill the trees in two or three years.

Treatment: When the blotches are about one-third to one-half inch in diameter, spray with "Black-leaf 40," tobacco extract, 1 gallon in 800 gallons of water, adding 4 pounds of whale-oil soap to each 100 gallons.

GOOSEBERRY

Fruit-worm (Dakruma convolutella): Larva, about three-fourths inch long, greenish or yellowish, feeding in the berry, causing it to ripen prematurely.

Treatment: Destroy affected berries. Clean cultivation. Poultry.

GRAPE

Grape berry-worm (Polychrosis viteana): Larva, about one-fourth inch long, feeding in the berry, often securing three or four together by a web; two broods.

Remedy: Spray with arsenate of lead before blossoms open. Repeat after blooming and again in early July. Destroy wormy berries in August.

Grape-curculio (Craponius inaequalis): Larva, small, white, with a brownish head. Infests the grape in June and July, causing a little black hole in the skin and a discoloration of the berry immediately around it. The adult is a grayish brown snout-beetle, about one-tenth inch long.

Treatment: Spray with arsenate of lead while the beetles are feeding on the leaves. The beetle may be jarred down on sheets, as with the plum-curculio. Bagging the clusters.

Grape-slug or Saw-fly (Selandria vitis): Larva about one-half inch long, yellowish green with black points, feeding upon the leaves; two broods.

Remedies: Arsenicals. Hellebore.

Grape Root-worm (Fidia viticida): The small white grubs feed upon the roots, often killing the vines in a few years. The adults are small grayish-brown beetles that eat peculiar chain-like holes in the leaves during July and August. Cultivate thoroughly in June, especially close around the vines to kill the pupae in the soil. At the first appearance of the beetles spray the plants with arsenate of lead at the rate of 8 or 10 pounds to 100 gallons of water, to which should be added 1 gallon of molasses.

Grape-vine Flea-beetle (Graptodera chalybea): Beetle, of a blue metallic colour, about one-fourth inch long, feeding upon the buds and tender shoots in early spring.

Treatment: Arsenicals to kill the grubs on leaves during May and June. The beetle can be caught by jarring on bright days.

Grape-vine Root-borer (Memythrus polistiformis): Larva, one and one-half inch or less long, working in the roots.

Preventive: Thorough cultivation during June and July.

Treatment: Dig out the borers.

Grape-vine Sphinx (Ampelophaga myron): A large larva, two inches long when mature, green with yellow spots and stripes, bearing a horn at the posterior extremity, feeding upon the leaves, and nipping off the young clusters of grapes; two broods.

Treatment: Hand-picking. Arsenicals early in the season. There are other large sphinx caterpillars which feed upon the foliage of the vine and which are readily kept in check by hand-picking and spraying.

Phylloxera (Phylloxera vastatrix): A minute insect preying upon the roots, and in one form causing galls upon the leaves.

Preventive: As a rule this insect is not destructive to American species of vines. Grafting upon resistant stocks is the most reliable method of dealing with the insect yet known. This precaution is taken to a large extent in European countries, as the European vine is particularly subject to attack.

Remedies: There is no reliable and widely practicable remedy known. Burn affected leaves. Bisulphide of carbon poured in holes in the ground, which are quickly filled, is very effective. Carbolic acid and water used in the same way is also recommended. Flood the vineyard.

Leaf-hopper (Typhlocyba comes): These small yellowish leaf-hoppers, erroneously called " thrips," suck the sap from the underside of the leaves, causing them to turn brown and dry up.

Treatment: Spray the under side of the leaves very thoroughly with whale-oil soap, 1 pound in 10 gallons of water, or with " Black-leaf" tobacco extract, 1 gallon to 100 gallons of water; or 1 gallon " Black-leaf 40 " in 1000 gallons of water about July 1, to kill the young leaf-hoppers. When using tobacco extract add about 2 pounds whale-oil soap to each 50 gallons to make it spread and stick better. Repeat the application in a week or ten days. In houses, tobacco-smoke, pyrethrum poured upon coals held under the vines, syringing with tobacco-water or soap suds.

Rose-chafer (Macrodactylus subspinosus): The ungainly, long-legged, grayish beetles occur in sandy regions, and often swarm into vineyards and destroy the blossoms and foliage.

Treatment: At the first appearance of the beetles spray with arsenate of lead at the rate of 8 or 10 pounds to 100 gallons of water, to which should be added 1 gallon of molasses.

Hollyhock. Bug (Orthotylus delicatus): A small green bug, attacking the hollyhock with great damage.

Treatment: Kerosene emulsion. Tobacco extracts.

Lawns Ants (Formica sp.): Insects burrowing in the ground, forming " ant hills."

Remedy: A tablespoonful of bisulfid of carbon poured into holes six inches deep and a foot apart, the holes being immediately filled up.

LETTUCE

Aphis or Green-fly: A plant-louse on forced lettuce.

Preventive: Tobacco-dust applied on the soil and plants as soon as the aphis makes its appearance, or even before. Renew every two or three weeks if necessary. Fumigating with tobacco is the surest remedy.

Cabbage-looper (Autographa brassiae): Larva, somewhat over an inch long, pale green, with stripes of a lighter colour, feeding on leaves of many plants, as cabbage, celery, and endive.

Remedies: Pyrethrum diluted with not more than three times its bulk of flour. Kerosene emulsion. Hot water.

MELON

Melon-worm (Diaphania hyalinata): Larva, some over an inch long, yellowish green and slightly hairy, feeding on melon-leaves, and eating holes into melons, cucumbers, and squashes; two or more broods.

Remedies: Hellebore. Arsenicals early in the season.

MUSHROOM

Mushroom-fly: The maggot bores through the stems of the mushrooms before they are full grown.

Preventive: Keep the beds cool so that the fly cannot develop. When the fly is present, growing mushrooms in warm weather is usually abandonded.

ONION

Maggot (Pegomya cepetorum): Much like the Cabbage Maggot.

Remedies: Carbolic acid emulsion. Bisulfid of carbon.

Thrips (Thrips tabaci): Minute elongate yellowish insects that cause a wilting and dying of the tops.

Treatment: Clean culture, kerosene emulsion, tobacco extracts.

ORANGE AND LEMON.

Purple Scale (Lepidosaphes beckii): An elongate brownish purple scale resembling an oyster-shell in shape.

Treatment: Fumigation, using heavy dosage.

Red-scale (Aspidiotus aurantii): A nearly circular reddish or yellowish scale.

Treatment: Fumigation. Distillate.

Black-scale (Saissetia olece): A large soft-bodied dark brown or nearly black scale.

Treatment: Fumigation. Distillate.

Mealy-bug (Pseudococcus citri): A mealy white soft-bodied insect nearly one-fourth inch long, occurring in masses in the angles of the branches, axils of the leaves, and around the stem of the fruit.

Treatment: Fumigation. Destruction of all rubbish under the trees.

Red-spider (Tetranychus sexmaculatus): Minute greenish yellow mites found on the leaves.

Treatment: Dry sulfur, or sulfur and water used as a spray.

White-fly (Aleyrodes citri and A. nubifera): The immature stages are found on the underside of the leaves and are scale-like in form. The adults are minute white-winged flies.

Treatment: Fumigation. Fungous diseases,

Rust-mite (Phytoptus oleivorus): A minute mite, causing the rust on oranges and lemons.

Treatment: Sulfur, dry or as a spray.

Thrips (Euthrips citri): A minute, active, yellow insect that scars the fruit and curls and distorts the leaves.

Treatment: Make four applications of lime-sulfur (33° Beau-me), 1 gallon in 75 gallons of water, adding " Black-leaf 40 " tobacco extract at the rate of 1 part in 1800 parts of the dilute lime-sulfur, as follows:—

- *First:* Just after most of the petals have fallen from the blossoms.
- *Second:* Ten or fourteen days after the first.
- *Third:* From three to four weeks after the second.
- *Fourth:* In August or September, to protect later growths of foliage.

Parsley. Parsley-worm (Papilio asterias): Larva, inch and a half long, light yellow or greenish yellow with lines and spots; feeding upon leaves of parsley, celery, carrot, etc. When the worm is disturbed it ejects two yellow horns, with an offensive odor, from the anterior end.

Remedies: Hand-picking. Poultry are said to eat them sometimes. Upon parsnip, arsenicals. Parsnip.

Parsnip Web-worm (Depressaria heracliana): Larva, about a half inch long, feeding in the flower cluster and causing it to become contorted.

Treatment: Arsenicals, applied as soon as the young worms appear, and before the cluster becomes distorted. Burn the distorted umbels. Destroy all wild carrots.

PEA

Pea-weevil or Pea-bug (Bruchus pisi): A small brown-black beetle, living in peas over winter. The beetle escapes in fall and spring, and lays its eggs in young pea-pods, and the grubs live in the growing peas.

Treatment: Hold over infested seed for one year before planting. Late planting in some localities. Fumigation with carbon bisulfid.

Pea Aphis (Macrosiphum pisi): A rather large green plant-louse, often attacking peas in great numbers and causing enormous losses.

Treatment: Rotation of crops. Early planting. When peas are grown in rows, the brush-and-cultivator method may be used.

The plant-lice are brushed from the plants with pine boughs, and a cultivator follows stirring the soil. This operation should be performed while the sun is hot and the ground dry. Most of the lice will be killed before they can crawl back to the plants. Repeat every three to seven days.

PEACH

Black Aphis (Aphis persicce-niger): A small black or brown plant-louse which attacks the tops and roots of peach-trees. When upon the roots it is a very serious enemy, stunting the tree and perhaps killing it. Thrives in sandy lands.

Treatment: Kerosene emulsion. Tobacco decoction and extracts.

Katydid: This insect is often troublesome to the peach in the southern states in the early spring, eating the leaves and girdling young stems.

Remedy: Poisoned baits placed about the tree.

Green Peach-louse or Aphis (Myzus persicae): A small insect feeding upon the young leaves, causing them to curl and die.

Treatment: Lime-sulfur, kerosene emulsion, or tobacco decoction. After the buds open, either of the last two.

Peach-tree Borer (Sanninoidea exitiosa): A whitish larva, about three-fourths inch long when mature, boring into the crown and upper roots of the peach, causing gum to exude.

Remedies: Dig out the borers in June and mound up the trees. At the same time apply gas-tar or coal-tar to the trunk from the roots up to a foot or more above the surface of the ground.

Peach Twig-moth (Anarsia lineatella): The larva of a moth, a fourth inch long, boring in the ends of the shoots, and later in the season attacking the fruit. Several broods.

Remedy: Spray with lime-sulfur just after the buds swell. Spray trunks and larger branches in late spring to kill first brood pupae in the curls of bark.

Peach-tree Bark-beetle (Phlaeotribus liminaris): A dark brown beetle one-tenth inch in length burrowing under the bark.

Treatment: Burn all brush and worthless trees as soon as the infestation is observed. Keep the trees in healthy condition by thorough cultivation and the use of fertilizers.

Apply a thick whitewash to the trunk and branches three times a season; first, the last week of March; second, second week in July; third, first week in October.

Fruit-tree Black-beetle (Scolytus rugulosus): A small beetle similar to the last.

Treatment: Burn all brush and worthless trees as soon as the infestation is observed. Keep the trees in healthy condition by thorough cultivation and the use of fertilizers.

Plum-curculio (Conotrachelus nenuphar): In Missouri and Georgia this insect has been successfully controlled on peach by spraying with arsenate of lead, 4 pounds to 100 gallons of self-boiled lime-sulfur. Spray, first when the "husks" drop from the fruit; second, ten days or two weeks later. It is unsafe to spray peaches more than twice with arsenate of lead.

Red-legged Flea-beetle (Haltica rufipes): A flea-beetle feeding on the leaves of peach trees, often in great numbers.

Remedies: The insects fall at once upon being jarred, and sheets saturated with kerosene may be used upon which to catch them. Spray with arsenate of lead in self-boiled lime-sulfur.

Midge (Diplosis pyrivora): A minute mosquito-like fly; lays eggs in flower-buds when they begin to show white. These hatch into minute grubs which distort and discolor the fruit.

Remedies: Destroy the infested pears. Cultivate and plow in late summer and fall to destroy the pupae then in the ground.

Pear-leaf Blister (Eriophyes pyri): A minute mite which causes black blisters to appear upon the leaves. The mites collect under the bud-scales in winter.

Remedy: Lime-sulfur or miscible oil as a dormant spray.

Pear-tree Borer (Sesia pyri): A small whitish larva, feeding under the bark of the pear tree.

Remedy: Same as for round-headed apple-tree borer.

Pear-twig Beetle (Xyleborus pyri): Brownish or black beetle, one-tenth inch long, boring in twigs, producing effect much like pear-blight, and hence often known as " pear-blight beetle." It escapes from a minute perforation at base of bud; probably two broods.

Treatment: Burn twigs before the beetle escapes.

Pear Psylla (Psylla pyricola): These minute, yellowish, flat-bodied, sucking insects are often found working in the axils of the leaves and fruit early in the season. They develop into minute, cicada-like jumping-lice. The young psyllas secrete a large quantity of honey-dew, in which a peculiar black fungus grows, giving the bark a characteristic sooty appearance. There may be four broods annually, and the trees are often seriously injured.

Treatment: Clean culture; remove rough bark from trunks and larger limbs to discourage adults from hibernating on the trees, and spray with miscible oils while trees are dormant. Spray with lime-sulfur wash at strengths used to combat scale, just before leaves appear, to destroy eggs. After blossoms have dropped, spray with whale-oil soap, 1 pound to 5 or 7 gallons of water; kerosene emulsion diluted with 8 to 12 parts of water; or standardized tobacco decoctions at strengths recommended on containers. If psyllas are abundant, trees should be frequently sprayed.

Pear Thrips (Euthrips pyri): Minute insects, 1/20 inch in length, dark brown when adult, white with red eyes when young, that attack the opening bud and young fruits in early spring. They suck the sap from the tender growth, and the females lay eggs in the fruit stems, causing a loss of the crop. The nymphs hibernate in the ground a few inches from the surface.

Treatment: Thorough cultivation during October, November, and December (in California). Make two applications of " Black-leaf " tobacco extract, 1 gallon in 60 gallons of 2 per cent distillate oil emulsion, the first just as the fruit buds begin to open, the second just after the petals fall. In the East it may be controlled by timely applications of tobacco extract and whale-oil soap.

PECAN

Bud-moth (Proteopteryx deludana): A brownish caterpillar about one-half inch in length, feeding on the opening buds in early spring and on the underside of the leaves in summer.

Treatment: Arsenate of lead in summer to kill larvae of second brood. Lime-sulfur and arsenate of lead in dormant season just before buds open, to destroy hibernating larvae.

Case-bearer (Acrobasis nebulella): A small caterpillar living inside a case which it carries with it. It attacks the opening buds.

Treatment: Arsenate of lead as soon as the buds begin to open. Repeat if necessary.

Borer (Sesia scituta): A wood-boring caterpillar working in the sapwood.

Treatment: Digging out.

Twig-girdler (Oncideres cingulatus): A brownish-gray beetle, about one-half inch long, which girdles twigs in August and September. The female lays eggs above the girdle. The twigs soon fall.

Remedy: Burn the twigs, either cutting them off or gathering them when they fall.

Persimmon: White Peach-scale (Diaspis pentagond).

Remedy: Lime-sulfur when the trees are thoroughly dormant.

Twig-girdlers (Oncideres cingulatus texana): Dark gray long-horned beetles that girdle the twigs, causing them to drop.

Remedy: Pick up and burn fallen twigs in fall and winter.

PINEAPPLE

Katydid (Acanthacara similis): A large katydid which

attacks, among other plants, the leaves of the pineapple. *Remedy:* Arsenicals, before the plants are mature.

Mealy-bugs (several species): These mealy white insects attack the plant at the base of the leaves, usually underground.

Treatment: Set only clean plants, or dip them in resin wash or kerosene emulsion. In the field apply tobacco dust freely in the bud before the bloom begins to appear, or spray with kerosene emulsion.

Red-Spider (Stigmaeus floridanus): Minute mites occurring in great number at the base of the leaf, where they induce rot.

Treatment: Tobacco dust applied to bud.

PLUM

Curculio (Conotrachelus nenuphar): Larva, a whitish grub, feeding in the fruit.

Remedies: Arsenate of lead, 6 pounds to 100 gallons of water; apply as soon as the calyx falls, and repeat two or three times at intervals of about ten days. Jarring the beetles on sheets very early in the morning, beginning when trees are in flower, and continuing from four to six weeks, is probably the most sure procedure. There are various styles of sheets or receptacles for catching the insects as they fall from the tree. Clean culture.

Plum-gouger (Coccotorus prunicida): A small larva, feeding upon the kernel of the plum. The beetle bores a round hole in the plum instead of making a crescent mark, like the curculio.

Remedy: Catch the beetles over a curculio-catcher.

Scale (Lecanium corni): A large circular scale occurring on plum (and perhaps other) trees in New York.

Remedy: Thorough spraying with kerosene emulsion, one part to five of water, in the winter. More dilute emulsion or tobacco extracts in midsummer, when the young insects are on the leaves and young shoots.

POPLAR

Cottonwood Leaf-beetle (Lina scripta): A striped beetle

feeding on the leaves and shoots of poplars and willows. *Remedy:* Arsenicals.

Poplar Borer (Cryptorhynchus lapathi): A whitish grub burrowing in the wood.

Treatment: In nurseries spray thoroughly about the middle of July with arsenate of lead to kill the parent beetles.

POTATO

Colorado Potato-beetle (Leptinotarsa decemlineata): Beetle and larva feed upon the leaves.

Remedies: Arsenicals, either dry or in spray, about a third stronger than for fruits. Hand-picking the beetle.

Stalk-weevil (Trichobaris trinotata): A grub boring in the stalk of the potato near or just below the ground. Serious at the West and in some places eastward.

Remedy: Pull all infested vines as soon as they wilt, and spread them in the sun where the insects will be killed. Burn the vines as soon as the crop is harvested. Destroy all solanaceous weeds.

Flea-beetles (Halticini): Small, dark-colored jumping beetles that riddle the leaves with holes.

Preventive: Bordeaux mixture as applied for potato blight acts as a repellent.

Potato Tuber-worm (Phthorimaea operculella): A small caterpillar burrowing in the stems and tubers both in the field and in storage.

Preventives: Clean cultivation, sheep and hogs to destroy the small potatoes left in the field after digging. Crop rotation over a considerable area. On digging remove the potatoes at once to an uninfested storeroom. Do not leave them on the field over-night.

Privet or Prim. Privet Web-worm (Diaphania quadristigmalis): Small larva feeding in webs on the young shoots of the privet, appearing early in the season; two to four broods.

Remedies: Trim the hedge as soon as the worms appear, and burn the trimmings. Probably the arsenicals will prove useful.

Quince-curculio (Conotrachelus crataegi): This curculio

is somewhat larger than that infesting the plum, and differs in its life-history. The grubs leave the fruits in the fall, and enter the ground, where they hibernate and transform to adults the next May, June, or July, depending on the season.

When the adults appear, jar them from the tree on to sheets or curculio-catchers and destroy them. To determine when they appear, jar a few trees daily, beginning the latter part of May.

RADISH

Maggot (Pegomya brassicae): Treated the same as the Cabbage-maggot, which see.

RASPBERRY

Cane-borer (Oberea bimaculata): Beetle, black, small, and slim; making two girdles about an inch apart near the tip of the cane, in June, and laying an egg just above the lower girdle; the larva, attaining the length of nearly an inch, bores down the cane. Also in blackberry.

Remedy: As soon as the tip of the cane wilts, cut it off at the lower girdle and burn it.

Raspberry Root-borer (Bembecia marginata): Larva about one inch long, boring in the roots and the lower parts of the cane, remaining in the root over winter.

Remedy: Dig out the borers.

Raspberry Saw-fly (Morwphadnus rubi): Larva about three-fourths inch long, green, feeding upon the leaves. *Remedies:* Hellebore. Arsenicals, after fruiting.

Root Gall-fly (Rhodites radicum): A small larva which produces galls on the roots of the raspberry, blackberry, and rose, causing the bush to appear sickly, and eventually killing it. The swellings are probably often confounded with the nematode root-galls.

Remedy: There is no remedy except to destroy the galls; if plants are badly affected, they must be dug up and burned.

Snowy or Tree-cricket (Ecanthus niveus): Small and whitish cricket-like insect, puncturing canes for two or three inches, and depositing eggs in the punctures.

Remedy: Burn infested canes in winter or very early spring.

RHUBARB

Rhubarb-curculio (Lixus concavus): A grub three-fourths inch long, boring into the crown and roots. It also attacks wild docks.

Remedy: Burn all infested plants, and keep down the docks. Hand-picking.

Mealy-bug: Tobacco extracts. Syringe the plants in the morning, and two hours later syringe again with clean water.

Rose Leaf-hopper (Typhlocyba rosae): A very small hopper, white, often mistaken for thrips, living on the leaves of roses.

Remedies: Whale-oil soap. Kerosene. Kerosene emulsion. Dry pyrethrum blown on bushes when leaves are wet. Tobacco extracts.

SQUASH

Borer or Root-borer (Melittia satyriniformis): Soft, white, grub-like larva which bores inside the stem and causes rot to develop, killing the vine.

Preventives: Plant early squashes as traps. As soon as the early crop is gathered, burn the vines to destroy eggs and larvae of the borer. Fall harrowing of infested fields will help to expose the pupae to the elements.

Cut out borers whenever found. After the vines have grown to some length, cover some of the joints with earth, so that a new root system will develop to sustain the plant in case the main root is injured.

STRAWBERRY

Crown-borer (Tyloderma fragariae): White grub, one-fifth inch long, boring into the crown of the plant in midsummer. The mature insect is a curculio or weevil.

Preventives: Rotation of crops. Isolation of new beds from infested beds. Plant uninfested plants.

Leaf-roller (Ancylis comptana): Larva, less than one-half inch long, feeding on the leaves, and rolling them up in threads of silk; two broods.

Treatment: Turn under in the fall all old beds that have become worthless. Spray with arsenate of lead, 4 pounds in 100 gallons of water, after the eggs are laid but before the leaves are folded – the first half of May in the latitude of New Jersey.

Root-borer (Anarsia sp.): Larva, about one-half inch long, whitish, boring into the crown of the plant late in the season, and remaining in it over winter.

Remedy: Burn the plant.

Root-louse (Aphis forbesii): From July to the close of the season the lice appear in great numbers on the crowns and on the roots of the plants.

Remedies: Rotation in planting. Disinfect plants coming from infested patches by dipping the crowns and roots in kerosene emulsion, or tobacco extract. Fumigation.

Saw-fly (Emphytus maculatus): Larva, nearly three-fourths inch long, greenish, feeding upon the leaves; two broods. Remedies: Hellebore. Arsenicals for second brood.

Weevil (Anthonomus signatus): Beetle, one-eighth inch long, reddish black, feeding on flower-buds, particularly those of the polleniferous varieties.

Preventives: Plant principally pistillate varieties. Every fifth row should be of some profusely flowering staminate variety

Sugar Cane Insects to insure pollinization. Clean culture. Destroy all wild blackberry and raspberry vines in the vicinity.

Root-borer (Typophorus canellus): A whitish grub one-eighth inch in length, feeding on the roots. The parent beetle is brownish, and appears in great numbers in May.

Treatment: Arsenicals to kill the beetles. Plant new beds at a distance from old ones.

SUGAR-CANE

Stalk-borer (Diatraea saccharalis): This is the " cane-borer " of the South, and is a species of long standing in the southern United States. The insects attack corn and sugar-cane.

In corn-growing areas in the South, it is known as " the larger corn stalk-borer." The eggs of the insect are laid on the cane-leaves, and the caterpillar of the moth develops within the cane-stalk. Between the months of May and December, the complete development of the insect occupies a period of a little over thirty days, that is, a brood may be expected about every month.

Treatment: The control measures consist of the burning of the trash after harvest, fall planting where possible, not to intercrop cane with corn, not to plant corn or cane on windrowed areas, that is, areas on which cane has been windrowed for the spring plant, and to cover all seed cane well to prevent the emergence of moths which may have developed from " borers " planted in the seed cane.

Mealy-bug (Pseudococcus calceolariae): Common on sugar-cane in the southern parishes of Louisiana, and recorded further in the United States from Florida and California. Known in Louisiana as " pou-a-pouche." The insects occur in a mass about the roots and beneath the lower leaf-sheaths of the cane plant, and the mass is covered by a white mealy secretion. The mealy-bug hibernates on the roots of the stubble beneath the surface of the ground or on the stalks put down in windrow as seed for the spring plant. Brood follows brood throughout the summer months.

Treatment: Burning of trash after harvest, fall planting, and the selection of seed cane from non-infested areas are the main methods that may be employed in the control of this species.

Root-beetle (Ligyrus rugiceps): This insect occurs throughout the lower Mississippi valley and the southern states generally as far north as North Carolina. As the name implies, the beetle infests the roots of the cane plant. The insect hibernates in the advanced larval or the pupal stages, and the adult appears in the spring.

The injury to the cane is accomplished by the adult eating into the young shoots just below the surface of the ground. From this point the insect works downwards to the roots, where the eggs are laid. The larva develops about the roots.

In the case of young shoots the injury is sufficient to practically sever the shoot from the mother cane or stubble. This kills out the heart of the young plant, and unless the cane suckers well, the stand is seriously affected.

Treatment: If the stubble cane is off-barred in the spring and the soil kept away from the young cane as late as the conditions will allow, much injury from the root beetle will be avoided.

Frequent cultivation of the plant cane will disturb the beetles in the soil and lessen their chance of attacking the cane. No great amount of vegetable matter should be plowed under on those areas where the root beetle is abundant, since this favors the development of the larvae or " white grubs."

The headlands and ditch banks should be kept clear of grass, since the beetle develops in these situations bordering the cane-fields. In districts where freezing temperatures occur, late fall plowing will turn out many of the grubs, and they will perish from exposure. During an attack, it is often profitable to have children follow and collect the beetles behind the hoe gangs.

SWEET-POTATO

Saw-fly (Schizocerus ebnus and S. privatus): Small larva about one-fourth inch long, working upon the leaves. The fly is about the size of a house-fly.

Remedies: Hellebore and arsenicals.

Root-borer (Cylas formicarius): A whitish grub one-fourth inch in length, burrowing through the tubers.

Preventive: Burn infested tubers and the vines.

Tortoise beetles (Cassidini): Beetles of brilliant colors and their slug-like larvae which eat holes in the leaves of newly reset plants.

Treatment: Dip the plants in a strong solution of arsenate of lead before resetting. Spray once or twice later with the same. Rotation of crops. Destroy all bindweed and wild morning-glory plants.

Flea-beetle. (Chaetocnema confinis): Small, dark-colored beetles, which attack the plants soon after they are reset.

Treatment: Dip the plants in a strong solution of arsenate of lead before resetting. Spray once or twice later with the same. Rotation of crops. Destroy all bindweed and wild morning-glory plants.

Cutworms: Poisoned bait. Late planting. Keep the land free from weeds the previous fall.

TOBACCO

Flea-beetle (Epitrix parvula): Small beetles eating holes in the leaves in the seed beds.

Treatment: Cover the beds tightly with canvas, or spray thoroughly with arsenate of lead, one pound in 12 gallons of water.

Cutworms: Use poisoned bait. Sod land should be plowed in fall.

Flea-beetles, Grasshoppers, and Tree-crickets: Attacking the crop in the field, may be controlled by spraying with arsenate of lead, 1 pound in 16 gallons of water.

TOMATO

Fruit-worm (Heliothis obsoleta): Larva, one inch in length, pale green or dark brown, faintly striped, feeding upon the fruit. Also on corn and cotton.

Treatment: Hand-picking. Avoid planting close to corn or cotton, or after either of these crops or after peas or beans. Practice fall or winter plowing.

Tomato-worm (Phlegethontius sexta and P. quinquemaculata):

A very large green worm feeding upon the stems and leaves of the tomato and husk tomato. Seldom abundant enough to be very serious; kept in check by parasites.

Remedies: Hand-picking. Rotation of crops. Clean culture.

TURKEYS

Flea-beetles: Dip the young plants in a strong solution of arsenate of lead. Bordeaux mixture acts as a repellent.

Violet. Aphis: Fumigation when grown under glass.

Gall-fly (Contarinia violicola): The adult is a minute

mosquito-like fly. The whitish or yellowish maggot feeds in folds of the opening leaves, which become deformed, turn brown, and die.

Treatment: Fumigation is practically of no value. Thorough hand-picking as soon as any sign of injury is noticed. Do not let the pest become established in a house.

Red-spider (Tetranychus bimaculatus): Minute mites which cause the leaves to turn paler and become yellowish.

Treatment: On greenhouse violets there is nothing better than a stiff spray of clear water so applied as not to drench the beds. Repeat the spraying once or twice a week.

WHEAT

Hessian-fly (Mayetiola destructor): A small maggot infesting the plant between the leaf sheath and the stem. When full grown they transform to the puparium or " flaxseed " stage.

Preventives: Crop rotation, destruction of all volunteer wheat. Burning stubble where practicable.

Joint-worms (Isosoma spp.): Small yellowish larvae found in the straw, causing hard knots or galls.

Preventives: Crop rotation. Heavy use of fertilizer to give a rapid growth. Burning of stubble wherever practicable.

WILLOW

Willow-worm (Euvanessa antiopa): Larva, nearly two inches long, black, feeding upon leaves of willow, elm, and poplar; two broods.

Remedy: Arsenicals.

Chapter 12

Disease Control in the Landscape

GENERAL DISEASES FOUND ON ORNAMENTAL PLANTS

Leaf and Flower Diseases

- **Disease: Anthracnose Blight (fungi)**

Fig. Anthracose on Maple Leaf.

Description: Irregular tan or brown areas are found on the leaves, especially along the major veins of the leaf blade. The condition usually is noticed in late spring on lower parts of the plant. Deformation and twisting of leaves often result from infections that occurred during bud break and leaf expansion.

The blight is commonly found on ash, white oak, maple, and sycamore, particularly when such plants are growing in damp, cool places. Frost injury, herbicide damage, and leaf scorch are often mistaken as anthracnose infection. Examine affected leaves closely for fungal fruiting bodies.

Management Options: Rake and burn fallen leaves and plant debris. Fertilize and root-irrigate to maintain vigour and promote refoliation of affected areas of the plant. Avoid overhead irrigation on or near affected plants, especially at night or late in the day.

Prune away surrounding shade- producing vegetation to promote dry air movement over the affected plants. Thin plants if needed to improve air movement and speed drying time. If necessary, spray with a properly labeled fungicide during bud break in the spring.

- **Botrytis blight (fungus)**

Fig. Botrytis Blight on Geranium Leaf.

Description: Leaves develop brown, rotted areas. Stems and flower stalks may suddenly wilt and fall over. Flower buds may turn dark and fail to open. Opened flowers appear rotted and discolored and fall prematurely. During wet weather, affected areas may be covered with brown, fuzzy sporulation of the Botrytis fungus.

Management Options: Cut off, rake, and burn plant debris in the fall. Avoid overhead irrigation late in the day or at night. Prune away shade-producing vegetation. Space plants well when planting. Thin plants if needed to improve air movement and speed drying time. If necessary, spray with a properly labeled material during cool, wet weather.

- **Powdery Mildew (fungi)**

Description: White, powdery areas appear on leaves, stems, flowers, or buds. If the infection is widespread, the entire leaf blade may be covered with a thin, white fungal, powdery growth. Some mildew fungi affect older leaves first, such as on oaks, lilacs, zinnias, or chrysanthemums. Others affect newer shoots, such as on roses, flowering dogwood, London plane tree, or crabapples. When new shoots are affected, leaf curling and shoot stunting and twisting may be severe.

Fig. Powdery Mildew on Azalea.

Management Options: Plant susceptible plants in a sunny location and do not crowd them. Prune away surrounding shade-producing vegetation. Thin plants if needed to improve air movement and speed drying time. Avoid overhead irrigation late in the day or at night. At the first sign of disease, spray regularly with a properly labeled fungicide. Note: Many powdery mildews are not serious enough to warrant the use of fungicide.

- **Sooty mold on leaves (fungi)**

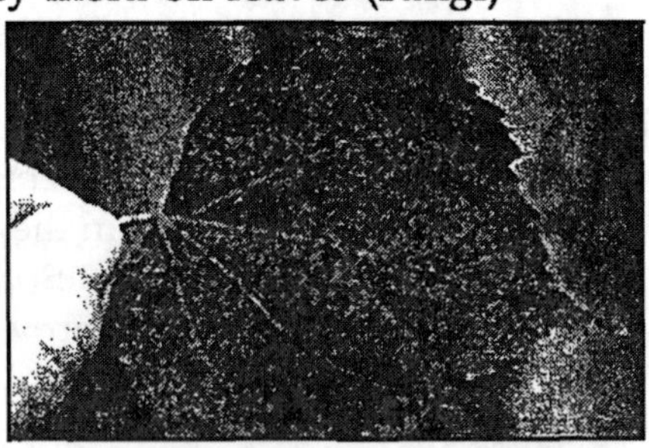

Fig. Sooty Mold on Linden Leaf

Description: Black sooty growth that partly or completely covers needles or leaves of a tree is caused by various fungi growing on honeydew secretion of sucking insects such as scales, aphids, and mealy bugs.

Management Options: Determine the insect problem, then use approved insecticides to control them. As the food supply is depleted, sooty mold growth will disappear. Wash plants with a weak soap solution if desired.

- **Rust (fungi)**

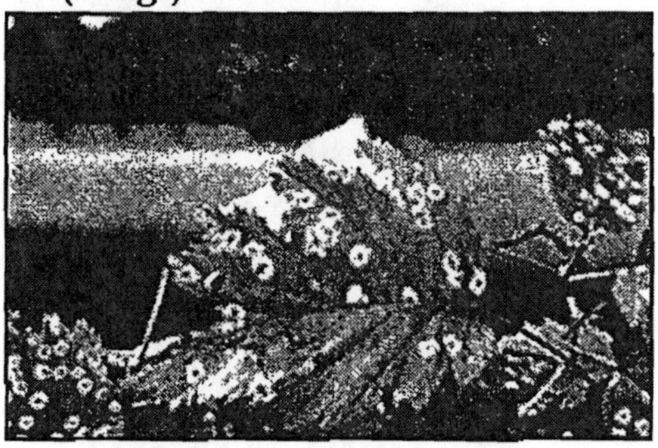

Fig. Cedar-hawthorn Rust on Hawthorn Leaf

Description: Orange to reddish-brown, often powdery pustules or spots will form on leaves, twigs, or fruit. Sometimes galling and disfigurement will be associated with the infections. In midsummer, infections on some plants will produce cup-like structures or tubular growths that will bear abundant orange spores.

Management Options: Resistant cultivars are available in some cases. Avoid overhead irrigation late in the day or night. Prune away shade-producing vegetation. Thin plants if needed to improve air movement and speed drying time. Rake and burn fallen leaves and plant debris. Rust infection may occur several days or weeks before symptoms are seen.

- **Dodder (Cuscuta) (parasitic seed plant)**

Description: Yellowish/orange vining plant strands grow, twist around, and quickly engulf infected host plants. Tiny flower and seed clusters are borne irregularly along the dodder vines. Affected plants eventually wither and die—mostly a problem on herbaceous ornamentals.

Management Options: Dodder is a parasitic flowering plant. It is an extremely infectious pest that is hard to eradicate once it becomes established. It has a wide host range. Remove and destroy all infected plant material at the first sign of an infestation. Apply a properly labeled pre-emergent herbicide regularly to affected areas for the next three or four growing seasons to destroy the persistent dodder seeds as they germinate.

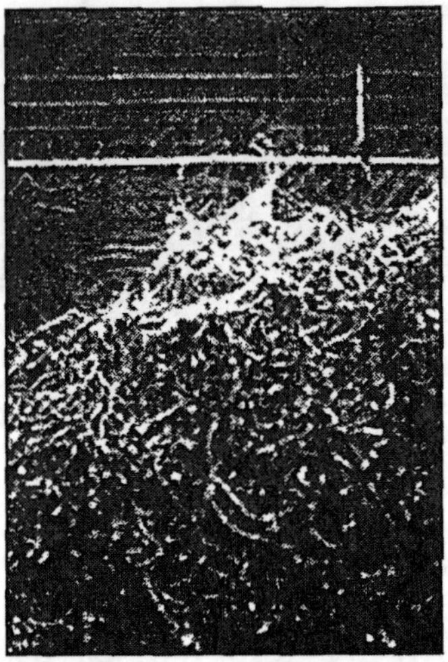

Fig. Dodder Infecting and Growing over the Surface of a Shrub

- **Leaf spot (fungi and bacteria)**

Fig. Phyllosticta or Purple-eye Leaf Spot on Maple

Description: Most leaf spots start on lower leaves. Many gradually increase in size to form large, dark brown to black areas. Others remain small and distinct but increase in numbers and begin to develop on younger leaves as the season progresses. Small, black, spore-forming pustules can often be seen in the spots when fungi are the cause.

Management Options: Avoid overhead irrigation, especially late in the day. Prune away surrounding shade-producing vegetation. Do not crowd newly planted material. Renovate older landscapes to reduce shade and plant crowding. For fungal leaf spots, spray plants at first sign of disease with a properly labeled fungicide. More than one spray may be needed. Leaf spot diseases of trees in the landscape are generally not damaging enough to warrant chemical control.

Twig and Branch Diseases

- **Cankers and twig or tip blights (fungi)**

Description: Canker infections show up as definite areas that vary in colour from the surrounding healthy bark. When cankers girdle the twig or branch, the end of the branch dies. Then the result is twig blight. Small bumps (spore-forming pustules of the fungus) are often seen in the canker. Under some conditions, these may produce an ooze of orange, white, or brownish spore masses during rainy weather. Close examination of blighted twigs will generally reveal areas of infection at the base of the dead tissue.

Management Options: Prune out cankers or blighted branches. Do not prune during warm, rainy weather. Reshaping of the plant may be necessary if many branches are removed. Fertilize to encourage new growth and quick healing of pruning wounds. Water during dry weather. Fungicide treatments are not effective preventives or curatives for cankers but may aid in prevention of twig or tip blights.

- **Fire blight (bacterium)**

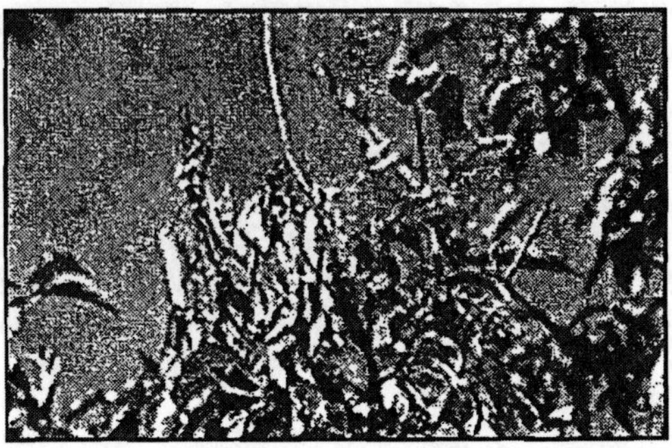

Fig. Fire Blight on Crabapple.

Description: Young twigs and branches die from the terminals and appear burned. Branches may be bent over at the tip, resembling a shepherd's crook. Dead leaves generally remain on the branch. Affects members of the rose family.

Management Options: Prune out branches well below signs of damage. Dip pruning tool in 70 per cent alcohol between each cut. Avoid heavy nitrogen fertilization, which produces more susceptible soft growth. Avoid splashing water. Plant resistant varieties or species when available.

Root and Crown Diseases

- **Chlorosis (or yellowing)**

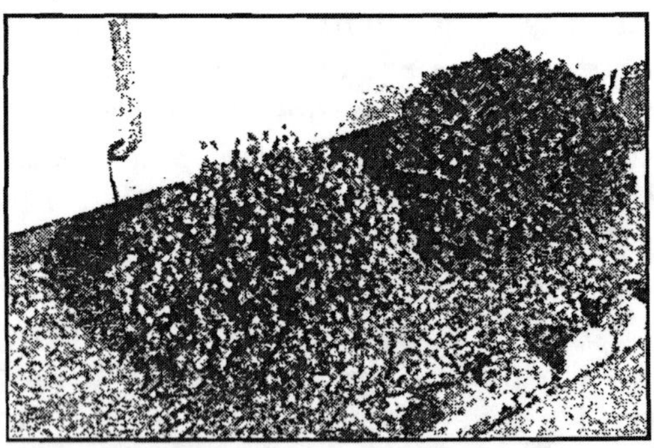

Description: This condition can occur in response to many types of environmental stresses, including those that result in root decline. Leaves will be smaller than normal. The yellowing may be general or only between the major veins of the leaf. One branch or the entire plant may be affected. In later stages or during droughts, leaf scorch, wilt, and leaf drop may occur.

Management Options: If the condition has resulted from a nutritional imbalance, fertilization with "trace" minerals and/or addition of materials to increase soil acidity (lower the pH) may help. Vertically mulch to relieve soil compaction if needed.

Water deeply once a week during dry spells, but only if the soil has dried to moderate levels within the fibrous root-growth areas. Control root-damaging insects with appropriate soil insecticides. Check for spider mites and treat appropriately. Improve drainage if wet soils are suspected to be the cause of the problem. Recently planted material can be lifted and replanted if planted too deeply.

- **Leaf scorch and dieback (decline)**

Fig. Physiological Leaf Scorch on Maple

Description: These diseases commonly result from winter damage as well as from poor root function coupled with high temperatures and moisture shortage. In some cases, bacterial infection of the plant is involved.

Scorched leaves are brown around the edges and sometimes between the veins. Leaf yellowing and wilting may occur in late summer. Dieback and decline is often mistaken for Verticillium wilt. No wood streaking is present, however. Girdling roots may also cause these symptoms.

Management Options: Investigate and attempt to correct the root stresses if present. Vertically mulch to open the soil, increase aeration, improve water penetration, and improve drainage. Water deeply once a week during dry spells. Pruning may be helpful to improve the root-shoot balance. Add plant fertilizers only if soil tests indicate shortages, imbalances, or pH problems.

Protect cold-sensitive plants from winter wind. Be sure soil is moist in the fall. Plants, especially evergreens, need moisture from the soil to counter the damaging effects of winter wind and sun. If girdling roots are the cause, it may be possible to cut them out.

- **Root, stem or crown rot (fungi)**

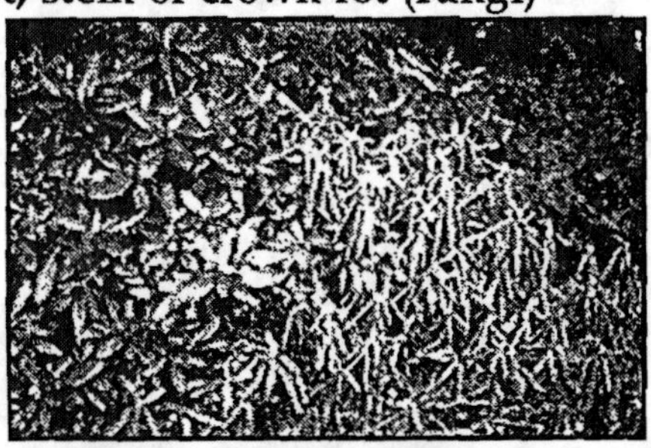

Fig. Above Ground Symptoms of Phytophrhora Root and Crown Rot on Rhododendron

Description: Leaves turn yellow, wilt, and droop. Leaf petioles and the base of the leaf blade may turn brown or black. Later, brown, rotted tissue may be evident at the base of the plant. Plants become loose and wobbly, pulling easily from the soil. These diseases are most severe during high soil moisture and high temperature periods.

Management Options: Once plants become infected, they rarely recover. Remove diseased plants. Improve drainage of the planting site. Do not plant sensitive plants in wet areas or cover roots with heavy mulches. Never use plastic sheeting as a mulch or weed preventer.

Porous landscape cloth or organic mulches are preferable to plastic. Buy disease-free plants with vigorous root systems. Before replanting affected areas, treat soil with a fungicide for water mold root rots and one for other root and stem-rotting fungi.

Vascular System Diseases

- **Crown gall (bacterium)**

Fig. Bacterial Crown on Gall Euonymus

Description: Infected plants may be stunted and do not flower well. Galls or larger masses of tissue form at the crown or at the point of grafting. Galls on stems and roots may be found on some plants.

Management Options: Purchase gall-free plants. Remove and burn severely infected plants. Keep plants vigorous with proper amounts of fertilizer and water. Control insect and mite pests that further weaken infected plants. Consider using other plants that are not susceptible.

- **Wetwood or slime flux (bacterium and yeast)**

Description: Sap seeps out of the infected areas through cracks, wounds, and pruning cuts and flows down the trunk. The foul-smelling substance often kills the bark. Insects are commonly attracted to the fermenting sap. This disease is often found on maple, birch, elm, oak, poplar, sycamore, and willow.

Management Options: There is no cure or preventive measure. Avoid damaging bark in any way.

Drill into the infected wood to relieve internal sap pressure. Holes should be 6 to 14 inches below the fluxing area, drilled on an upward slant. Insert a piece of pipe to carry the flux away from the tree and base of the trunk. Water infected trees deeply once a week during dry spells.

- **Verticillium wilt and dieback (fungus)**

Description: Wilting and yellowing of foliage is followed by premature defoliation. One limb of the tree may be affected first. Gradual dieback may be apparent. Yellow-brown, brown, black, or greenish-black streaks may be found in outer rings of wood of infected branches. In herbaceous plants, the vascular system of stems and roots are discolored.

Fig. Above-ground Symptoms of Verticium Wilt on Maple

Management Options: If herbaceous ornamentals are involved, resistant cultivars may be available. If resistant cultivars are not available, do not plant susceptible plants in infested beds. Severely infected trees will have to be removed. Trees with one infected branch should have the branch removed and the tree fertilized and watered during dry spells. Apply mulch after planting. Use resistant plant types if replanting is needed.

- **Bacterial leaf scorch (bacteria)**

Description: These diseases commonly result from winter damage as well as from poor root function coupled with high temperatures and moisture shortage. In some cases, bacterial infection of the plant is involved.

Scorched leaves are brown around the edges and sometimes between the veins. Leaf yellowing and wilting may occur in late summer. Dieback and decline is often mistaken for Verticillium wilt.

No wood streaking is present, however. Girdling roots may also cause these symptoms. Rapid dieback may be due to damage from rabbits, mice or moles feeding on the lower stems or trunks.

Management Options: If bacterial leaf scorch is suspected, consult the plant and pest diagnostic clinic or a professional arborist for diagnosis.

SPECIFIC DISEASES OF TREES, SHRUBS AND GROUND COVERS

Ash (Fraxinus)

- **Anthracnose (fungus)**

Fig. Anthracnose on Ash Leaves

Description: Irregular tan or brown areas are found on the leaves, especially along the major veins of the leaf blade. The condition usually is noticed in late spring on lower parts of the plant. Deformation and twisting of leaves often result from infections.

Management Options: Control in landscape usually not necessary. Green ash reported to be less susceptible.

- **Heartwood rot (fungus)**

Description: On large or mature trees large, woody, perennial, hoof-shaped mushrooms or conks occur on the trunks.

Management Options: This fungus decays the non-living

heartwood. This fungus invades trees through wounds, often those caused by storm damage or improper pruning. Fungicides are not effective.

Ajuga

- **Crown and root rot (fungus)**

Description: Wilting and discoloration of lower leaves, soon death of plant. Roots are rotted. Strands or mats of white fungus at base of plant.

Very small, round, seed-like bodies form on and around the plant. These bodies are light colored at first, changing to buff and then brown.

Management Options: Usually develops under warm and wet conditions. Remove and destroy affected plants. A fungicide drench may be necessary before replanting.

Alpine Currant (Ribes)

- **Anthracnose leaf spot (fungus)**

Description: Small, brown spots develop first on the lower leaves. Leaves yellow and fall.

Management Options: Destroy fallen diseased leaves when possible. Fungicides may be necessary. Worse during wet weather or when overhead watering is used.

Arborvitae (Thuja)

- **Fall needle browning (physiological)**

Fig. Normal Fall Browning of Older Foliage on Arborvitae

Description: Foliage on inner branchlets turn yellow or brown.

Management Options: A normal process. No control necessary.

Azalea (Rhododendron)

- **Leaf spot (fungus)**

Description: Most ash leaf spots are circular and light to dark brown. White, powdery areas appear on leaves, stems, flowers, or buds. If the infection **is** widespread, the entire leaf blade may be covered with a **thin**, white fungal, powdery growth. Some mildew fungi affect older leaves first, such as on oaks, lilacs, zinnias.

Management Options: Seldom serious on the landscape; control not necessary.

- **Yellows (mycoplasma)**

Description: A general yellowing of the tree is associated with production of shoots from buds that form along the stem. Leaves on these shoots are often small and misshapen. Growth reduced.

Management Options: No specific controls are available o feasible. Green ash reported to be tolerant.

- **Leaf spot (fungus)**

Description: Angular to irregular brown spots. Lower leaves usually affected first. Some defoliation may occur.

Management Options: Some cultivars are very susceptible. This disease is most frequent during wet seasons or when plants were watered overhead. More often seen in greenhouses. Fungicides may be used.

- **Leaf gall (fungus)**

Fig. Exobasidium Leaf gall on Azalea.

Description: Conspicuous white galls on leaves and stem.

Management Options: An oddity; does little damage. Pick off the galls.

- **Powdery mildew (fungus)**

Fig. Powdery Mildew on Azalea

Description: White, powdery areas appear on leaves, stems, flowers, or buds. If the infection is widespread, the entire leaf blade may be covered with a thin, white fungal, powdery growth. Some mildew fungi affect older leaves first, such as on oaks, lilacs, zinnias.

Management Options: Deciduous azaleas vary from susceptible to highly resistant.

Barberry (Berberis)

- **Verticillium wilt (fungus)**

Description: Wilting and yellowing of foliage is followed by premature defoliation. One limb of the tree may be affected first. Gradual dieback may be apparent. Yellow-brown, brown, black, or greenish-black streaks may be found in outer rings of wood of infected branches. In herbaceous plants, the vascular system of stems and roots are discolored.

Beech (Fagus)

- **Heartwood rots (fungi)**

Description: On older or mature trees, decay of the heartwood of trunk and limbs is common.

Management Options: Avoid injury or wounds to the

bark. Follow accepted pruning practices. Repair wounds due to storm damage. Fungicides are not effective.

Birch (Betula)

- **Cankers (fungi)**

Description: Young twigs and branches die from the terminals and appear burned. Branches may be bent over at the tip, resembling a shepherd's crook. Dead leaves generally remain on the branch. Affects members of the rose family.

Management Options: Not usually a problem on river birch, unless under extreme drought or low fertility.

- **Iron chlorosis (physiological)**

Description: On high pH soils interveinal chlorosis may develop due to unavailability of iron.

- **Leaf spot (fungus)**

Description: Anthracnose leaf spot on some species, especially river birch. Small, round spots with a dark border.

Management Options: Control usually not necessary.

- **Rust (fungus)**

Description: White, powdery areas appear on leaves, stems, flowers, or buds. If the infection is widespread, the entire leaf blade may be covered with a thin, white fungal, powdery growth. Some mildew fungi affect older leaves first, such as on oaks, lilacs, zinnias.

Management Options: Not a problem on river birch, the most commonly planted species. Occasionally reported on less frequently planted species.

Black locust (Robinia)

- **Trunk and limb decay or rot (fungi)**

Description: At first there may be no outward signs of decay or rot. However, if cavities or areas of exposed wood occur on the trunk, or if branch stubs are deteriorating, wood rot is likely. Decay is certain if mushrooms in the form of bracket or conks are seen on the limbs or trunk. This is a problem of older or mature trees.

Management Options: The most effective procedure against this problem is prevention. Decay fungi enter only through wounds. If wounds occur they need to be treated immediately. If pruning is necessary follow accepted methods—no stubs. Fungicides are not effective.

Boxwood (Buxus)

- **Cankers and leaf spot (fungi)**

Description: Poor growth occurs on some branches. Leaves die. Small, rose-colored, waxy pustules are commonly found on leaves and branches. These pustules are generally caused by secondary fungi colonizing stressed tissues. Another fungus is found on yellow leaves forming small, black fruiting bodies.

Management Options: Remove and destroy infected parts. Remove leaves from branch crotches. Found mostly on winter-injured cultivars or plants stressed for other reasons. Plant only those cultivars known to be winter hardy.

Buckthorn (Rhamnus)

- **Decline (nematodes)**

Description: Stunting and yellowing is caused by root-feeding nematodes. This is the principal cause of decline of buckthorn.

Management Options: Do not replant in affected areas.

Callery pear (Pyrus)

- **Bacterial fire blight (bacteria)**

Description: Young twigs and branches die from the terminals and appear burned. Branches may be bent over at the tip, resembling a shepherd's crook. Dead leaves generally remain on the branch. Affects members of the rose family.

Management Options: Some cultivars are reported resistant. Bradford cultivar not recommended. Fire blight is not as severe on Callery pear as on certain highly susceptible apples, pears, crabapples, and firethorn. Infection is generally limited to short shoots.

Catalpa (Catalpa)

- **Leaf spots (fungi)**

Description: Brown spots scattered over the leaf. Worse on trees injured by midges, according to some reports.

Management Options: Control midges with appropriate insecticides. Leaf spots are generally not serious enough to warrant specific control measures.

- **Verticillium wilt (fungus)**

Description: Wilting and yellowing of foliage is followed by premature defoliation.

One limb of the tree may be affected first. Gradual dieback may be apparent. Yellow-brown, brown, black, or greenish-black streaks may be found in outer rings of wood of infected branches. In herbaceous plants, the vascular system of stems and roots are discolored.

Cinquefoil, Shrub or Shrubby (Potentilla)

- **Powdery mildew (fungus)**

Description: White, powdery areas appear on leaves, stems, flowers, or buds. If the infection is widespread, the entire leaf blade may be covered with a thin, white fungal, powdery growth. Some mildew fungi affect older leaves first, such as on oaks, lilacs, zinnias.

Management Options: Seldom require control measures.

- **Rust (fungus):** White, powdery areas appear on leaves, stems, flowers, or buds. If the infection is widespread, the entire leaf blade may be covered with a thin, white fungal, powdery growth. Some mildew fungi affect older leaves first, such as on oaks, lilacs, zinnias.

Cotoneaster (Cotoneaster)

- **Canker (fungus)**

Description: Young twigs and branches die from the terminals and appear burned. Branches may be bent over at the tip, resembling a shepherd's crook. Dead leaves generally remain on the branch. Affects members of the rose family. Dieback of branches on some species, especially following severe winters.

Management Options: Infection usually occurs after

plants are predisposed by winter injury. Prune out affected branches.

- **Fireblight (bacterium)**

Description: Young twigs and branches die from the terminals and appear burned. Branches may be bent over at the tip, resembling a shepherd's crook. Dead leaves generally remain on the branch. Affects members of the rose family.

Management Options: Some species are resistant.

Crabapple (Malus)

- **Fireblight (bacterium)**

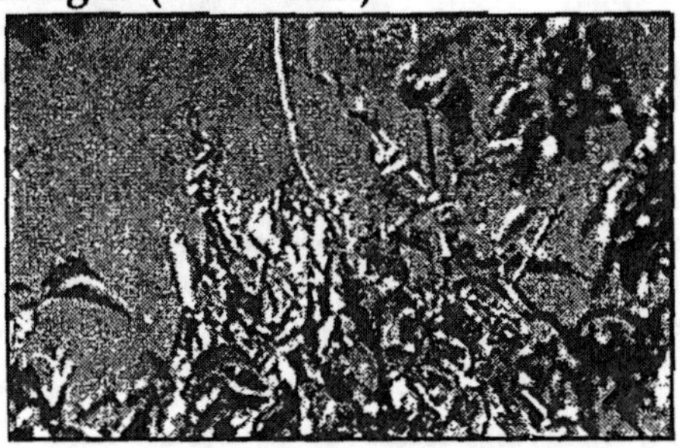

Fig. Fire Blight on Crabapple

Description: Sudden wilting, dying, and browning or blackening of new shoots. Leaves on these shoots die; ends of branches appear burned. Initial infections often occur on blossoms.

Management Options: Remove infected branches by pruning them off 8 to 12 inches below the diseased tissue. Avoid excess nitrogen fertilizer. Plant resistant varieties. If pruning shears are used, wipe them well with 70 per cent alcohol between each cut.

- **Frogeye leaf spot (fungus)**

Description: Tan spots are small but can run together to cause large dead areas. White, powdery areas appear on leaves, stems, flowers, or buds.

If the infection is widespread, the entire leaf blade may be covered with a thin, white fungal, powdery growth. Some mildew fungi affect older leaves first, such as on oaks, lilacs, zinnias.

Management Options: Not common or serious on most recommended varieties

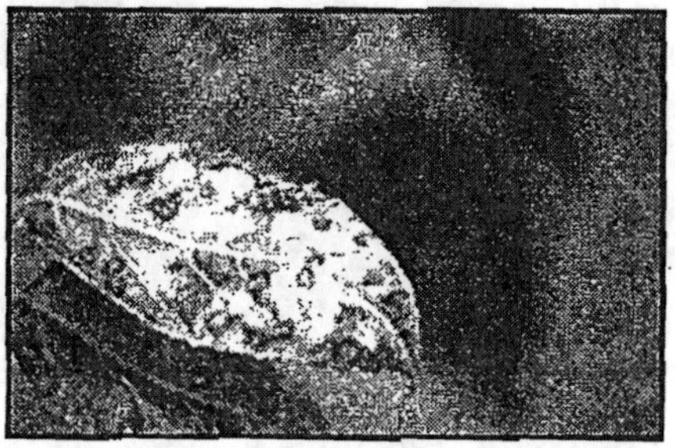

Fig. Frogeye Leaf Spot on Crabapple

- **Powdery mildew (fungus)**

Description: White, powdery areas appear on leaves, stems, flowers, or buds. If the infection is widespread, the entire leaf blade may be covered with a thin, white fungal, powdery growth. Some mildew fungi affect older leaves first, such as on oaks, lilacs, zinnias.

Management Options: Not common or serious on recommended varieties.

- **Rust (fungus)**

Description: White, powdery areas appear on leaves, stems, flowers, or buds.

If the infection is widespread, the entire leaf blade may be covered with a thin, white fungal, powdery growth. Some mildew fungi affect older leaves first, such as on oaks, lilacs, zinnias, The fungus that causes this disease also produces rust galls on some species of junipers.

Management Options: Not common or serious on most recommended varieties.

- **Scab (fungus)**

Description: Olive green to black spots with radiating edges on leaves. Leaves turn yellow and fall prematurely beginning in June or July. Small, round dark areas that become corky are found on the fruits. Trees may be 75 per cent defoliated 6 to 8 weeks before normal time.

Management Options: Rake and destroy or haul away fallen leaves in the fall. Plant resistant varieties.

Fig. Apple Scab on Crabapple Leaves

- **Scorch and dieback**

Description: These diseases commonly result from winter damage as well as from poor root function coupled with high temperatures and moisture shortage. In some cases, bacterial infection of the plant is involved. Scorched leaves are brown around the edges and sometimes between the veins. Leaf yellowing and wilting may occur in late summer. Dieback and decline is often mistaken for Verticillium wilt. No wood streaking is present, however. Girdling roots may also cause these symptoms.

Dogwood (Cornus)

- **Anthracnose blight (fungus)**

Fig. Anthracose on Dogwoods Leaves

Description: Small, purple-bordered spots and irregular blotches on leaves. Flower bracts can also be affected. Twigs can be infected through leaf petioles or through small lesions in the bark of shoots. Twig dieback occurs and in time larger branches die. The leaf spot phase can be mistaken for other fungal leaf spot diseases. If we suspect dogwood anthracnose, send samples to the Plant Pest and Diagnostic Laboratory.

Management Options: Flowering dogwood is very susceptible, although some cultivars may be less susceptible. Other dogwoods are more resistant. Kousa or Chinese dogwood is sometimes used as replacement, although it flowers a month later. Avoid heavily shaded sites. Fungicides applied at bloom and one month later may be helpful. Some Cornus florida X Cornus kousa hybrids are reported resistant.

- **Canker and dieback (fungus)**

Description: Dieback of branches and stems. Cankers form and contain small raised black dots which contain the fungus spores. Prevalent on the shrubby species of dogwood.

Management Options: Water during dry periods. Fertilize properly. Prune out diseased branches.

- **Leaf spot (fungi)**

Description: Angular, grayish spots with dark purple margins found first on lower leaves, especially during wet weather. Septoria fungi are the most common causes.

Management Options: Red and yellow twigged dogwoods are particularly susceptible. Avoid overhead watering. Fungicides may be used if necessary.

- **Powdery mildew (fungus)**

Fig. Powdery Mildew on Dogwood

Description: White, powdery areas appear on leaves, stems, flowers, or buds. If the infection is widespread, the entire leaf blade may be covered with a thin, white fungal, powdery growth. Some mildew fungi affect older leaves first, such as on oaks, lilacs, zinnias, Becoming a more common problem. New growth is usually affected with leaves twisted and rolled. Reddening of foliage is sometimes apparent. On some species found on older leaves.

Management Options: Plant in lightly shaded area. Avoid dense shade. Promote conditions that improve air circulation.

- **Scorch and dieback (physiological)**

Description: These diseases commonly result from winter damage as well as from poor root function coupled with high temperatures and moisture shortage. In some cases, bacterial infection of the plant is involved. Scorched leaves are brown around the edges and sometimes between the veins. Leaf yellowing and wilting may occur in late summer. Dieback and decline is often mistaken for Verticillium wilt. No wood streaking is present, however. Girdling roots may also cause these symptoms.

Management Options: Do not plant in full sun or other unfavorable sites. Control borers.

Elms (Ulmus)

- **Black leaf spot (fungus)**

Description: The most common leaf spot is small, black, slightly raised and becomes shiny. Spots may be numerous. Leaves turn yellow and fall.

Management Options: Most frequent on Chinese and American elm. Control not usually necessary. Destroy fallen leaves, if feasible.

- **Elm yellows or Elm phloem necrosis (mycoplasma)**

Description: This resembles Dutch elm disease in its early stages. Wilting and curling of still- green leaves is also a common early symptom. Leaves eventually turn yellow, then brown.

These early symptoms are more apt to be throughout the tree, unlike the early symptoms of Dutch elm disease. Discoloration of cambium and phloem tissue (inner bark) at the base of the tree is perhaps the most distinctive symptom. The tissue turns yellow or caramel brown. It often has a wintergreen odor. Death of the tree is fairly rapid.

Management Options: Control of the elm leafhopper, which carries the infectious agent, should be carried out as a protective measure if the disease is present on other trees in the area. This may be difficult. European and Asiatic elms and some hybrids are resistant. Diseased trees should be removed.

- **Wilt and dieback or Dutch elm disease (fungus)**

Fig. Dutch Elm Disease

Fig. Insect Galleries Left by the European Elm Bark Beetle

Description: Yellowing, wilting, and then browning will appear on leaves of individual branches. The sapwood in these branches will be streaked dark brown.

The streaks will show up as circles or rings of brown in

cross section. The fungus spreads rapidly in most elms, eventually killing the tree.

Management Options: Integrated control programs can be effective in slowing the progress of the disease in a community. These programs involve controlling the insect vector (European elm bark beetle) of the fungus with proper insecticide treatment, prompt and complete pruning out of infections as soon as identified, removal and destruction of dead trees in the area, root graft isolation of healthy trees, use of resistant elm hybrids or varieties, and injection treatment with fungicides. Trained arborists should be consulted regarding the implementation of these integrated control programs. A community-wide effort generally is necessary. Plant hybrids and selections tolerant or resistant; be sure they are resistant to the elm leaf beetle, a serious insect pest.

- **Scorch and dieback (physiological)**

Description: These diseases commonly result from winter damage as well as from poor root function coupled with high temperatures and moisture shortage. In some cases, bacterial infection of the plant is involved. Scorched leaves are brown around the edges and sometimes between the veins. Leaf yellowing and wilting may occur in late summer. Dieback and decline is often mistaken for Verticillium wilt. No wood streaking is present, however. Girdling roots may also cause these symptoms.

- **Wetwood or slime flux (fungus)**

Description: Wilting and yellowing of foliage is followed by premature defoliation. One limb of the tree may be affected first. Gradual dieback may be apparent. Yellow-brown, brown, black, or greenish-black streaks may be found in outer rings of wood of infected branches. In herbaceous plants, the vascular system of stems and roots are discolored.

Euonymus (Euonymus)

- **Anthracnose (fungus)**

Description: Irregular tan or brown areas are found on the leaves, especially along the major veins of the leaf blade. The condition usually is noticed in late spring on lower parts of the plant. Deformation and twisting of leaves often result from infections. One of several leaf spot diseases, none of which are routinely common or serious.

Management Options: Fungicides can be used, but usually not necessary in landscape situations.

- Crown gall (bacterium)

Fig. Bacterial Crown Gall on Euonymus

Description: Roots and stems are infected with large galls. Generally a problem on evergreen types.

Management Options: Destroy heavily infected plants. Prune out galls if few are present. Disinfest pruning shears between cuts in 70 per cent alcohol. Plant other shrub types if problem persists. E. alatus (burning bush) is not susceptible.

- **Powdery mildew (fungus)**

Description: White, powdery areas appear on leaves, stems, flowers, or buds. If the infection is widespread, the entire leaf blade may be covered with a thin, white fungal, powdery growth. Some mildew fungi affect older leaves first, such as on oaks, lilacs, zinnias.

Management Options: A problem on some evergreen types, not a problem on burning bush. Chemical sprays not usually necessary.

Firethorn (Pyracantha)

- **Fire blight (bacterium)**

Description: Young twigs and branches die from the terminals and appear burned. Branches may be bent over at the tip, resembling a shepherd's crook. Dead leaves generally remain on the branch. Affects members of the rose family. Blossoms may be blackened. Shoots dieback. Leaves do not drop. ***Management Options:*** Do not purchase infected plants. Some varieties are resistant.

- Scab (fungus)

Fig. Scab on Pyrocantha Fruit

Description: Fruits are covered with scabby lesions and turn black. Leaves are covered with dark sooty areas, turn yellow and drop.

Management Options: Plant resistant varieties. If susceptible, spray before and after flowering with a fungicide.

Forsythia (Forsythia)

- **Description: Failure to bloom (abiotic)**

Description: Flowers fail to appear on otherwise healthy plants.

Management Options: The plant forms flower buds for the following year in the late summer. Because forsythia is such a vigorous grower, many gardeners prune the plant back during the winter. This removes the flower buds. In addition, low winter temperatures (below -10oF) often kill the flower buds.

- **Twig and branch gall (fungus)**

Description: Nodular, abnormal growths of varying sizes appear along stems. In severe cases, twigs and branches will die.

Management Options: This disease was once thought to be due to the crown gall bacterium. Recent evidence points to a fungal pathogen in some cases and a genetic disorder in other cases. Prune out and destroy infected branches. No chemical sprays are known to be effective.

- **Twig blight, dieback (fungus)**

Description: Caused by Sclerotinia and recognized by black, hard, conspicuous bodies in the pith when stems are split length-wise. These fungal structures are white at first.

Management Options: Prune and destroy affected branches. Not a common problem.

Hawthorn (Crataegus)

- **Fire blight (bacterium)**

Description: Young twigs and branches die from the terminals and appear burned. Branches may be bent over at the tip, resembling a shepherd's crook. Dead leaves generally remain on the branch. Affects members of the rose family.

Management Options: Many species or cultivars are resistant or only slightly susceptible.

- **Leaf spot (fungus)**

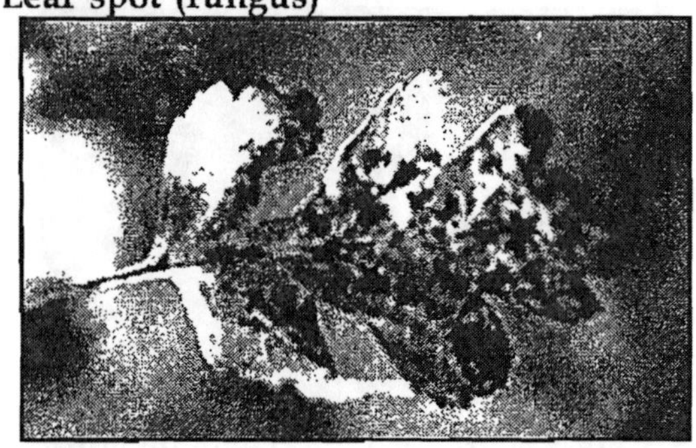

Fig. Entomosporium Leaf Spot on Hawthorn Leaf.

Description: English hawthorns are susceptible to a summer leaf spot that can lead to extensive defoliation. The disease begins as a tiny purple spotting.

Leaves yellow and drop as the disease develops on younger foliage. White, powdery areas appear on leaves, stems, flowers, or buds.

If the infection is widespread, the entire leaf blade may be covered with a thin, white fungal, powdery growth. Some mildew fungi affect older leaves first, such as on oaks, lilacs, zinnias.

Management Options: Plant resistant or immune species or cultivars. Chemical control often necessary for susceptible English hawthorns if they are planted. Most hawthorns are not susceptible.

- **Rust on leaves, fruit and stems (fungi)**

Fig. Cedar-hawthorn Rust on Hawthorn Leaf

Fig. Cedar-quince on Hawthorn Fruit

Description: The most important rust species on hawthorns in the landscape affects twigs, stems, leaf petioles, and fruits. These plant parts become distorted and by summer an abundance of orange-yellow spores form, especially on the fruit. Leaf rust may also occur in some cases. Twig dieback can occur. Washington hawthorn is especially susceptible.

Management Options: The spores from the hawthorns spread the rust to susceptible junipers, the alternate hosts. Plant resistant hawthorns or do not plant near susceptible junipers.

Fungicides can be used beginning as the leaves unfold and continuing through the flowering period.

Hemlock (Tsuga)

- **Canker (fungi)**

Description: Young twigs and branches die from the terminals and appear burned. Branches may be bent over at the tip, resembling a shepherd's crook. Dead leaves generally remain on the branch. Affects members of the rose family.

Management Options: Most often affects plants in an unfavorable planting site such as exposure to full sun. Keep plants mulched, and water during dry periods; avoid injuries.

- **Needle browning (physiological)**

Description: Salt spray often results in damage. Some browning of older needles is a normal process in the fall.

Management Options: Do not plant near streets or roads where salt is commonly used.

- **Rust (fungus)**

Description: Hemlock rust commonly occurs on young shoot tips, where it causes twisting and deformation. White, powdery areas appear on leaves, stems, flowers, or buds. If the infection is widespread, the entire leaf blade may be covered with a thin, white fungal, powdery growth. Some mildew fungi affect older leaves first, such as on oaks, lilacs, zinnias.

Management Options: Usually not serious or life threatening.

Hickory (Carya)

- **Anthracnose (fungus)**

Description: Irregular tan or brown areas are found on

the leaves, especially along the major veins of the leaf blade. The condition usually is noticed in late spring on lower parts of the plant. Deformation and twisting of leaves often result from infections.

Management Options: Control is seldom necessary.

- **Leaf spot (fungus)**

Description: Several leaf spots may occur on hickory. None are usually serious or they occur late in the season.

Management Options: Not usually serious enough to warrant control measures.

- **Scorch and dieback (physiological)**

Description: These diseases commonly result from winter damage as well as from poor root function coupled with high temperatures and moisture shortage. In some cases, bacterial infection of the plant is involved. Scorched leaves are brown around the edges and sometimes between the veins. Leaf yellowing and wilting may occur in late summer. Dieback and decline is often mistaken for Verticillium wilt. No wood streaking is present, however. Girdling roots may also cause these symptoms.

Holly (Ilex)

- **Black root rot (fungus)**

Description: Plants are chlorotic and may be stunted, eventually with dieback. Roots are stunted and blackened. Inspect hollies in containers before purchase.

Management Options: Remove diseased plants once disease is confirmed.

- **Purple leaf spots (fungus)**

Description: Large, sometimes irregular spots occur on leaves. The spots may be yellow, brown, or black. Do not confuse with spine spot. White, powdery areas appear on leaves, stems, flowers, or buds. If the infection is widespread, the entire leaf blade may be covered with a thin, white fungal, powdery growth. Some mildew fungi affect older leaves first, such as on oaks, lilacs, zinnias.

Management Options: Several kinds of leaf spots are reported; control is usually not necessary. Not common.

- **Scorch and dieback (physiological)**

Description: These diseases commonly result from winter damage as well as from poor root function coupled with high temperatures and moisture shortage. In some cases, bacterial infection of the plant is involved. Scorched leaves are brown around the edges and sometimes between the veins. Leaf yellowing and wilting may occur in late summer. Dieback and decline is often mistaken for Verticillium wilt. No wood streaking is present, however. Girdling roots may also cause these symptoms.

- **Spine spot (insects or mechanical)**

Description: Small, gray spots, sometimes with purple, halos are found on the leaves. Once said to be caused by spines of adjacent leaves, but is caused by feeding of the holly leaf miner insect.

Management Options: Control of insects with appropriate sprays may be necessary.

Honey locust (Gleditsia)

- **Ganoderma rot (fungus)**

Description: A decay or rot of lower trunk and buttress roots occurs on older or mature trees. There may be no outward symptoms at first. Later large, shelf mushroom fruiting bodies grow at the base of the tree. They are cream-colored on the lower surface and brownish-red and shiny on top.

Management Options: Preventing lawn-mower wounds or any other injury to the bark and roots is the key to preventing this disease.

If wounds occur proper treatment immediately is necessary, but not always successful.

Honeysuckle (Lonicera)

- **Leaf blight (fungus)**

Description: Large areas of leaves turn dark brown and curl, and there is a whitish growth visible on the lower leaf surface. The disease is favored by cool, wet weather.

Management Options: Avoid overhead irrigation. Prune

away shade-producing vegetation. If necessary, spray with a properly labeled fungicide, although seldom necessary.

- **Powdery mildew (fungus)**

Description: White, powdery areas appear on leaves, stems, flowers, or buds. If the infection is widespread, the entire leaf blade may be covered with a thin, white fungal, powdery growth. Some mildew fungi affect older leaves first, such as on oaks, lilacs, zinnias.

Management Options: Not common or serious.

Horsechestnut and Buckeye (Aesculus)

- **Leaf blotch (fungus)**

Description: Small to large irregular, reddish-brown spots with yellowish edges occur on leaves in late June or July. Spots enlarge with browning and curling of leaflets. The disease is more prevalent on lower leaves.

Management Options: Plant in open sunny areas to increase air movement and leaf drying. Chemical control not usually necessary. Bottlebrush buckeye appears to be resistant.

- **Powdery mildew (fungus)**

Description: White, powdery areas appear on leaves, stems, flowers, or buds. If the infection is widespread, the entire leaf blade may be covered with a thin, white fungal, powdery growth. Some mildew fungi affect older leaves first, such as on oaks, lilacs, zinnias.

Management Options: No control recommended.

- **Scorch**

Description: These diseases commonly result from winter

damage as well as from poor root function coupled with high temperatures and moisture shortage. In some cases, bacterial infection of the plant is involved. Scorched leaves are brown around the edges and sometimes between the veins. Leaf yellowing and wilting may occur in late summer. Dieback and decline is often mistaken for Verticillium wilt. No wood streaking is present, however. Girdling roots may also cause these symptoms.

Management Options: Often confused with leaf blotch and vice-versa. Often present by late summer, especially on buckeye. Watering may delay onset of scorch.

Hydrangea (Hydrangea)

- **Foliar diseases (fungus)**

Description: Leaf spot, powdery mildew and leaf rust may occur occasionally on some species, but usually not serious.

Management Options: Control usually not necessary.

Ivy, Boston and Virginia creeper (Parthenocissus)

- **Leaf spot (fungus)**

Description: Small, circular, light tan spots develop on plants in cool, damp areas. Blighting and leaf deformation may occur later. White, powdery areas appear on leaves, stems, flowers, or buds.

If the infection is widespread, the entire leaf blade may be covered with a thin, white fungal, powdery growth. Some mildew fungi affect older leaves first, such as on oaks, lilacs, zinnias.

Management Options: Generally not serious enough to warrant control.

- **Powdery mildew (fungus)**

Description: White, powdery areas appear on leaves, stems, flowers, or buds. If the infection is widespread, the entire leaf blade may be covered with a thin, white fungal, powdery growth. Some mildew fungi affect older leaves first, such as on oaks, lilacs, zinnias, On Virginia creeper, but not serious.

Management Options: Does not warrant control.

Ivy, English (Hedera)

- **Bacterial leaf spot and stem canker (bacterium)**

Fig. Bacterial Leaf Spot on English Ivy.

Description: Leaf spots are brown or black and often appear water-soaked with reddish margins. Leaf stalks become black and shriveled. Canker forms in the woody portion of the vine.

Management Options: Remove infected plant material. Avoid splashing water and watering later in the day. If necessary, spray with a properly labeled bacteriacide.

Japanese Kerria (Kerria)

- **Leaf and twig blight (fungus)**

Description: Round to angular brown spots, with darker borders. Leaves yellow and die. Stem cankers develop and cause dieback of stem tips.

Management Options: Prune out diseased tissue. Fungicides may be necessary in severe cases.

Juniper (Juniperus)

- **Dieback (physiological)**

Description: Needles and small branches turn brown and die. Larger branches and sometimes an entire side may be involved or the entire plant dies.

Management Options: Avoid planting in dry, hot locations or in heavily shaded sites. Cover ground around plants with mulch to conserve water and control soil moisture variation. Fall watering is important. However, excess water may drown the roots.

- **Fall browning (physiological)**

Description: Some interior and older needles brown and fall in autumn, as a normal process.

Management Options: Be sure the plants are well watered and mulched, otherwise no control is needed.

- **Phomopsis tip blight (fungus)**

Fig. Over-all Appearance of a Juniper Affected with Phomopsis Tip Blight.

Description: Damages new growth and succulent branch tips of junipers from mid-April through September. Older, mature foliage is resistant to infection; therefore, most blighting occurs on the terminal 4 to 6 inches of the branches. Affected foliage first turns dull red or brown and finally ash-gray. Small gray lesions often girdle branch tips and cause blighting of foliage beyond the diseased tissue.

Small, black, spore-containing fungal fruiting bodies develop near the base of dead tips. Use a hand lens to view these diagnostic fungal structures more easily. Most infections usually occur in April through early June and again in late August through September. Very few infections occur in mid-summer or during the winter months. Repeated blighting in early summer can result in abnormal bunching (witches' broom) and discoloration of the foliage, stunting of young plants, or in severe cases, plant death.

Management Options: Management options for Phomopsis and Kabatina tip blights are essentially the same for both diseases; except for fungicide controls. When purchasing new plants, select those that have been reported to have disease resistance. Space new plantings to provide good ventilation and air circulation and avoid heavily shaded areas.

Avoid wounding plants, especially in spring and fall. Water plants in early morning so the foliage will dry as soon as possible. Maintain adequate fertility, but do not over-fertilize. Prune out diseased branch tips during dry summer weather and destroy them. Do so only when plants are dry and no rain or overhead irrigation is expected for several days. Avoid excessive pruning or shearing.

Chemical control of these tip blight diseases normally is not necessary in established landscape or windbreak plantings. Occasionally, fungicide applications may be needed on susceptible junipers to control Phomopsis blight. Application of certain copper-based (Phyton-27, Kocide) or thiophanate (Cleary's 3336, Topsin-M) fungicides at 7 to 21 day intervals during rapid plant growth in the spring will give adequate control of Phomopsis but not Kabatina tip blight.

Kabatina blight infections occur in the fall, and there currently are no fungicides labeled for control of this disease. These fungal tip blights are often mistaken for damage due to mites or insects.

- **Kabatina tip blight (fungus)**

Description: Symptoms first appear in February and March and well before those of Phomopsis tip blight. The terminal 2 to 6 inches of diseased branches throughout the juniper first turn dull green, then red or yellow. Small ash gray to silver lesions dotted with tiny, black, fruiting bodies of the fungus are visible at the base of the discolored tissue. The brown, desiccated foliage eventually drops from the tree in late May or June.

Foliar blighting occurs only in early spring; it does not continue through the summer. Blighting is also restricted to the branch tips and does not cause extensive branch dieback or tree death.

The primary infection period for the Kabatina fungus is thought to be in autumn, even though visible symptoms are not apparent until late winter or early spring. Infection often is associated with small wounds on branch tips caused by insect feeding or mechanical damage.

- **Rust Cedar—apple rust Cedar—hawthorn rust (fungi)**

Fig. Cedar-apple Rust Gall on Juniper

Fig. Cedar-apple Rust Gall on Juniper after the Gelatinous Spore Horns are Formed

Description: Dark brown, corky, spherical to kidney-shaped galls are produced on the leaves and stems of Eastern red cedar and some other species of Juniperus. During spring rains, the galls develop elongated, orange, gelatinous projections called spore horns. These spore horns produce spores that cause infections on alternate hosts. Alternate hosts are crabapples and hawthorns.

Management Options: Hand pick galls in early April if only a few are present. Do not plant susceptible junipers near susceptible apples, crabapples, or hawthorns.

- **Cedar-quince rust (fungus)**

Fig. Cedar-quince Rust on Juniper Twig

Description: Slight swellings on stems and twigs of susceptible junipers. The affected twigs are hardly noticed until yellow-orange, jelly-like spore masses form in late April and May. The twigs may die back.

Management Options: The most important alternate hosts are hawthorns, especially "Washington." Most other species are not generally severely infected. Do not plant susceptible junipers and hawthorns in the same area.

- **Salt burn (chemical)**

Description: Browning of branches occurs on side toward roadway where drift blows salty water onto plants or where salt-laden snow is shoveled onto plants.

Management Options: Prune out dead material in spring. Use protective barriers in problem areas, or use salt-tolerant plants. Some junipers are tolerant or resistant.

Leucothoe (Leucothoe)

- Leaf spots (fungi)

Description: Several species of fungi cause leaf spots that may vary in colour and size.

Management Options: Fungicides may be necessary in some cases. On isolated plants pick off and destroy affected leaves.

Lilac (Syringa)

- **Bacterial blight (bacterium)**

Description: Young shoots are marked by black stripes. One side of the shoot may be black. Water-soaked spots appear on leaves. Flower buds may be completely blackened. The problem is serious in wet weather and may be confused with frost injury.

Management Options: Prune out infected branches. Avoid excessive fertilization. Thin plants to improve air circulation. Prune away shade-producing vegetation. Chemical sprays are not effective.

- **Leaf roll and scorch**

Description: These diseases commonly result from winter damage as well as from poor root function coupled with high temperatures and moisture shortage. In some cases, bacterial infection of the plant is involved. Scorched leaves are brown around the edges and sometimes between the veins. Leaf yellowing and wilting may occur in late summer. Dieback and decline is often mistaken for Verticillium wilt. No wood streaking is present, however. Girdling roots may also cause these symptoms.

- **Leaf spot and blight (fungus)**

Description: Large brown spots with rings occur in damp weather. As they age, they commonly crack in the centre. White, powdery areas appear on leaves, stems, flowers, or buds. If the infection is widespread, the entire leaf blade may be covered with a thin, white fungal, powdery growth. Some mildew fungi affect older leaves first, such as on oaks, lilacs, zinnias.

Management Options: Seldom a common or serious problem.

- **Powdery mildew (fungus)**

Description: White, powdery areas appear on leaves, stems, flowers, or buds. If the infection is widespread, the entire leaf blade may be covered with a thin, white fungal, powdery growth. Some mildew fungi affect older leaves first, such as on oaks, lilacs, zinnias.

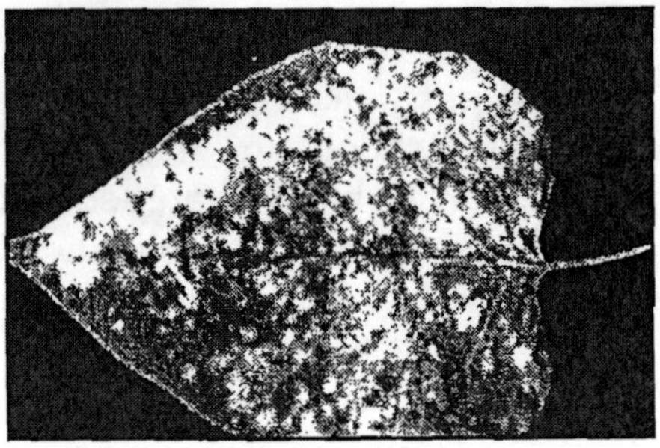

Fig. Powdery Mildew on Lilac

Management Options: Common lilac most susceptible.

Magnolia (Magnolia)

- **Frost injury**

Description: Flower buds or flowers are browned and killed when temperatures below freezing occur.

Management Options: Early-flowering species such as star magnolia and saucer magnolia are most commonly affected.

- Leaf spot (fungus or bacterium)

Description: Large, irregular spots form on older leaves in damp weather. White, powdery areas appear on leaves, stems, flowers, or buds. If the infection is widespread, the entire leaf blade may be covered with a thin, white fungal, powdery growth. Some mildew fungi affect older leaves first, such as on oaks, lilacs, zinnias.

Management Options: Fungal leaf spot is seldom a serious problem. Bacterial leaf spot has the potential to be serious, especially in nursery situations with overhead watering. Any practice directed toward maintaining dry foliage is beneficial.

Maple (Acer)

- **Anthracnose blight (fungus)**

Description: On Norway maple, narrow, irregular, brown streaks of dead tissue along the veins sometimes develop. On sugar, red, and other maples, large, irregular brownish areas between and along veins occur with symptoms similar to physiological leaf scorch.

Small fruiting bodies on and near the veins sometimes can be seen with a hand lens. Some defoliation may occur. Irregular tan or brown areas are found on the leaves, especially along the major veins of the leaf blade. The condition usually is noticed in late spring on lower parts of the plant. Deformation and twisting of leaves often result from infections.

Management Options: Control with fungicides usually not necessary. It may be helpful to rake and destroy fallen leaves to reduce over wintering inoculum of the fungus. NOTE: Black maple is often affected with another type of anthracnose. Circular tan to brown spots are scattered over the leaf surface. A different fungus is involved.

- **Chlorosis or yellowing**

Description: Leaves on one or several branches develop yellowing between the veins.

The veins remain green. In other cases, entire leaves yellow due to various physiological stresses. Young twigs and branches die from the terminals and appear burned. Branches may be bent over at the tip, resembling a shepherd's crook. Dead leaves generally remain on the branch. Affects members of the rose family.

Management Options: Interveinal chlorosis of red, silver, sugar, and Norway maples is often due to nutritional deficiencies as a result of soil pH above 6.5. At high soil pH, manganese becomes unavailable.

Lowering soil pH (often difficult), spraying foliage with micronutrients, or using trunk implants of manganese materials is sometimes recommended, especially for red maple.

- **Phyllosticta leaf spot (fungus)**

Description: Several fungi cause leaf spots on maple. The most common is Phyllosticta leaf spot. A light tan spot develops up to 8mm in diameter. Tiny black fungal spore bodies dot the surface.

Management Options: Leaf spots on maples are seldom severe enough to warrant chemical control.

Fig. Phyllosticta or Purple-eye Leaf Spot on Maple.

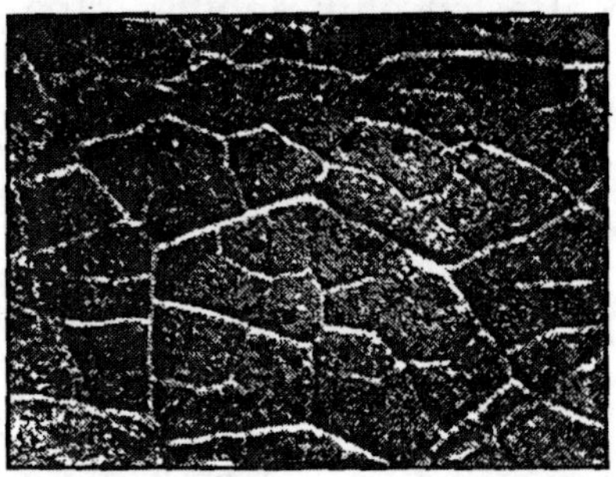

Fig. Close-up of Phyllosticta Leaf Spot on Maple

- **Tar spot (fungus)**

Fig. Tar Spot on Maple Leaves

Description: Another spot, called tar spot appears as shiny black spots up to.5 to 2 cm across. They look like spots of dried tar.

Management Options: Control seldom necessary. Generally not damaging and occurs late in the season.

- **Powdery mildew (fungus)**

Description: White, powdery areas appear on leaves, stems, flowers, or buds. If the infection is widespread, the entire leaf blade may be covered with a thin, white fungal, powdery growth.

Management Options: Some mildew fungi affect older leaves first, such as on oaks, lilacs, zinnias.

108. Scorch and dieback (physiological)

Fig. Physiological Leaf Scorch on Maple

Description: These diseases commonly result from winter damage as well as from poor root function coupled with high temperatures and moisture shortage. In some cases, bacterial infection of the plant is involved. Scorched leaves are brown around the edges and sometimes between the veins. Leaf yellowing and wilting may occur in late summer. Dieback and decline is often mistaken for Verticillium wilt. No wood streaking is present, however. Girdling roots may also cause these symptoms.

- **Verticillium wilt (fungus)**

Description: Wilting and yellowing of foliage is followed by premature defoliation. One limb of the tree may be affected first. Gradual dieback may be apparent.

Yellow-brown, brown, black, or greenish-black streaks may be found in outer rings of wood of infected branches. In herbaceous plants, the vascular system of stems and roots are discolored.

Management Options: Red maple less susceptible than Norway, sugar, and silver.

Fig. Above Ground Symptoms of Verticullium Wilt on Maple

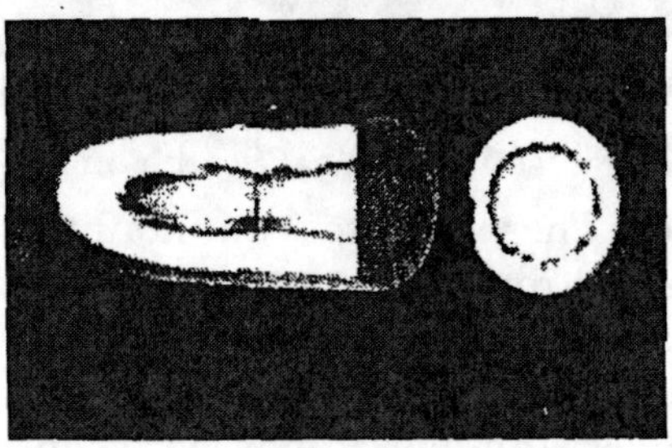

Fig. Cross-section and Longitudinal Section Through a Male Twig from a Tree Affected by Vertillium Wilt

Mock orange (Philadelphus)

- **Bacterial leaf spot (bacterium)**

Description: White, powdery areas appear on leaves, stems, flowers, or buds. If the infection is widespread, the entire leaf blade may be covered with a thin, white fungal, powdery growth. Some mildew fungi affect older leaves first, such as on oaks, lilacs, zinnias.

Management Options: Avoid overhead watering; not common in the landscape.

- **Powdery mildew (fungus)**

Description: White, powdery areas appear on leaves, stems, flowers, or buds. If the infection is widespread, the entire leaf blade may be covered with a thin, white fungal, powdery growth. Some mildew fungi affect older leaves first, such as on oaks, lilacs, zinnias.

Management Options: Controls not usually necessary.

Mountain Ash (Sorbus)

- **Cankers (fungi)**

Description: Young twigs and branches die from the terminals and appear burned. Branches may be bent over at the tip, resembling a shepherd's crook. Dead leaves generally remain on the branch. Affects members of the rose family. Almost always develops on stressed trees (recent transplants, drought, winter injury, etc.)

- **Fire blight (bacterium)**

Description: Young twigs and branches die from the terminals and appear burned. Branches may be bent over at the tip, resembling a shepherd's crook. Dead leaves generally remain on the branch. Affects members of the rose family.

- **Scorch and canker complex (fungi and environmental causes)**

Description: This is a major cause of mountain ash decline. Sunscald injury to the bark (southwest and west side) or any other injury to the bark provides entry points for one or more fungi. As cankers develop, yellowing or browning of leaves in the crown or on individual branches occur.

Management Options: Any protection from the sun of the thin bark on the south and west side will be helpful, and care to avoid all other types of injury to the bark is suggested.

Oak (Quercus)

- **Anthracnose (fungus)**

Description: Irregular tan or brown areas are found on the leaves, especially along the major veins of the leaf blade.

The condition usually is noticed in late spring on lower parts of the plant. Deformation and twisting of leaves often result from infections that occ Most common on white oak, but also on other species.

Management Options: Rarely life threatening; control not necessary.

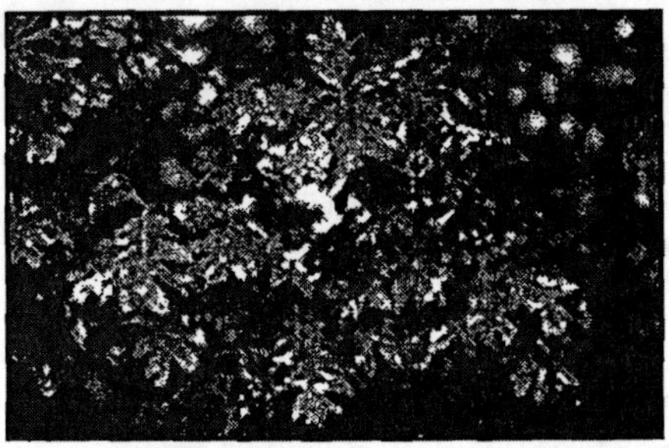

Fig. Anthracnose on Oak Leaves

- **Bacterial leaf scorch (bacterium)**

Description: Browning of leaf margins progressing toward the middle of the leaf. At first a single branch may be affected, often near the top of the tree. Leaves at the base of the branch develop symptoms first, but soon all leaves on the branch or limb are affected. Over time the diseased tree becomes less vigorous, leaves drop, and dieback occurs. May be confused with physiological leaf scorch. Oaks are most susceptible, but it has been reported on red maple, elm, and sycamore.

Management Options: No feasible management available. Remove badly diseased trees. Keep trees as healthy as possible until disease becomes severe enough to remove them. This disease cannot be easily or positively diagnosed without special tests in the laboratory.

- **Chlorosis or yellowing (physiological)**

Description: Young twigs and branches die from the terminals and appear burned. Branches may be bent over at the tip, resembling a shepherd's crook. Dead leaves generally remain on the branch.

Affects members of the rose family. Symptoms similar to those described for maple, but iron deficiency instead of manganese deficiency is the usual cause. Most common on pin oak and white oak.

Management Options: Use iron trunk implants.

Fig. Interveinal Chlorosis or Yellowing on Pin Oak Leaves

- **Leaf spots (fungi)**

Description: Several different fungi cause spots of various sizes, usually round, often with a colored border.

Management Options: Control measures not usually necessasy. These leaf spots are not serious or life threating.

- **Leaf blister (fungus)**

Fig. Taphrina Leaf Blister on Oak Leaves

Description: Circular, raised, blister-like, green, yellow to white areas up to 12 inch in diameter appear on upper leaf surface. Disease occurs on "red oaks" and is more prevalent in wet years. "White oak" group is also susceptible, but less commonly affected.

Management Options: Not usually serious enough to warrant control.

- **Lower trunk and buttress root rots (fungi)**

Description: Common cause is a spongy yellow-orange shelf fungus growing at the base of the diseased tree. These lower trunk and root rots usually become apparent only on older trees and trees with injuries to the bark.

Management Options: Preventing wounds is the key management option. Any injury at base of tree, such as lawnmower wounds, is a likely point of invasion by these fungi.

- **Oak wilt (fungus)**

Description: Wilting of leaves develops, usually first in the upper crown. Leaves become pale green, followed by yellowing or bronzing. The discoloration spreads from the margin toward the leaf midrib. Symptoms progress downward until all foliage is affected. Trees may die within a few weeks. A dark discoloration develops in the sapwood of wilting branches. The disease can be confirmed by laboratory tests.

Management Options: Oaks of the red oak group are most susceptible. White oaks are less susceptible, with only a few branches dying each growing season. Removal of diseased trees is recommended. The fungus can spread from a diseased tree to an adjacent healthy oak through natural root grafts. Therefore, deep trenching between such trees is sometimes recommended. Once a tree is infected, control is not possible.

- Powdery mildew (fungus)

Fig. Powdery Mildew on Oak Leaves

Description: White, powdery areas appear on leaves, stems, flowers, or buds. If the infection is widespread, the entire leaf blade may be covered with a thin, white fungal, powdery growth. Some mildew fungi affect older leaves first, such as on oaks, lilacs, zinnias, Some cultivars of English oak are particularly susceptible.

Management Options: On most species, control is not necessary.

- **Scorch and dieback (physiological)**

Description: These diseases commonly result from winter damage as well as from poor root function coupled with high temperatures and moisture shortage. In some cases, bacterial infection of the plant is involved. Scorched leaves are brown around the edges and sometimes between the veins. Leaf yellowing and wilting may occur in late summer. Dieback and decline is often mistaken for Verticillium wilt. No wood streaking is present, however. Girdling roots may also cause these symptoms.

- **Twig cankers (fungus)**

Description: Young twigs and branches die from the terminals and appear burned. Branches may be bent over at the tip, resembling a shepherd's crook. Dead leaves generally remain on the branch. Affects members of the rose family.

- **Wetwood or slime flux (fungus)**

Description: Wilting and yellowing of foliage is followed by premature defoliation. One limb of the tree may be affected first. Gradual dieback may be apparent. Yellow-brown, brown, black, or greenish-black streaks may be found in outer rings of wood of infected branches. In herbaceous plants, the vascular system of stems and roots are discolored.

Pachysandra (Pachysandra)

- **Leaf and stem blight (fungus)**

Description: Leaves appear first with brown blotches and finally as a blight. A brown to blackish stem rot with orange-brown, visible spore pustules develop.

Management Options: Remove diseased plants. Improve air circulation by thinning plants. If necessary, spray with a properly labeled fungicide, beginning as soon as new growth appears in the spring.

Fig. Volutella Leaf and Stem Blight of Pachysandra

- **Leaf yellowing**

Description: Overall yellow coloration of foliage develops, often in sunny locations or in sites with high soil pH (greater than 7.0).

Management Options: Avoid planting in full sun location. Since pachysandra is an acid-loving plant, modify soil pH to about 6.5 if possible.

Periwinkle (Vinca)

- **Stem blight and dieback (fungus)**

Fig. Leaf and Stem Dieback of Vinca Minor

Description: Individual stems wilt and blacken. Dark brown to black leaf spots may occur.

Management Options: Avoid overhead irrigation. Spray with a properly labeled fungicide, beginning when new shoots appear in the spring.

- **Root rot (fungus)**

Description: These diseases commonly result from winter damage as well as from poor root function coupled with high temperatures and moisture shortage. In some cases, bacterial infection of the plant is involved. Scorched leaves are brown around the edges and sometimes between the veins. Leaf yellowing and wilting may occur in late summer. Dieback and decline is often mistaken for Verticillium wilt. No wood streaking is present, however. Girdling roots may also cause these symptoms.

Photinia, Oriental (Photinia)

- **Fire blight (bacterium)**

Description: Young twigs and branches die from the terminals and appear burned. Branches may be bent over at the tip, resembling a shepherd's crook. Dead leaves generally remain on the branch. Affects members of the rose family.

Pieris (Pieris) (Japanese Andromeda)

- **Chlorosis**

Description: Young twigs and branches die from the terminals and appear burned. Branches may be bent over at the tip, resembling a shepherd's crook. Dead leaves generally remain on the branch. Affects members of the rose family.

Management Options: Check soil pH, and if alkaline, modify to about 6.5 or lower. Improve drainage and maintain mulch.

- **Dieback (fungus)**

Fig. Phytophthora Stem Blight and Dieback of Pieris

Description: Individual branches dieback. Both dieback diseases described under rhododendron are potential problems of Pieris.

Pine (Pinus)

- **Canker (fungus)**

Description: Young twigs and branches die from the terminals and appear burned. Branches may be bent over at the tip, resembling a shepherd's crook. Dead leaves generally remain on the branch. Affects members of the rose family.

- **Chlorosis or yellowing**

Description: Young twigs and branches die from the terminals and appear burned. Branches may be bent over at the tip, resembling a shepherd's crook. Dead leaves generally remain on the branch. Affects members of the rose family. Needles light green to yellow with limited annual stem growth.

Management Options: Air pollution maybe the cause on white pine.

- **Needlecasts and needle blights (fungi)**

Fig. Lophodermium Needle Blight on Scots Pine

Description: Pines, primarily Scots, Austrian and red, are subject to needle infection by several fungi. The fungi cause various spots ranging from black to whitish. As the disease progresses, the needles brown and fall from the trees.

Management Options: Plant pines in open sites where they can get sun and moderately drying winds. Do not crowd plants. Spray during infection periods with properly labeled fungicides. Infection periods vary depending on the particular fungus involved. Thus, a specific diagnosis is necessary to correctly schedule fungicide applications. Fungicides usually not necessary in the landscape. Fungicides are important on Christmas tree farms.

- **Needle rust (fungus)**

Fig. Pine Needle Rust on Red Pine

Description: Small, cream-to orange-colored pustules on needles. The pustules later break open to reveal an abundance of orange, dusty spores. Red pine is very susceptible. The fungus is not found on five-needled (white) pines.

Management Options: Destroy the alternate hosts (goldenrod and wild asters) near valuable plantings of red pines. Plant pines in full sun where moderate air movement will dry plants off after dew or rains. Spray plants in the fall with a properly labeled fungicide. Seldom damaging except on very young trees. For Austrian pine, the alternate hosts are ironweeds.

- **Pine wilt (nematode)**

Description: Early symptoms are on individual branches; the needles droop, turn brown, and adhere. Normal resin flow from cut branches ceases. Usually the entire tree is killed within a few months. Laboratory examination is necessary to confirm the diagnosis.

Management Options: Increase the vigour of pines with

fertilization if needed. Apply deep root waterings during dry spells. Promptly remove and destroy all pines in the area that have died from this disease or that have been confirmed to be infected with the nematode. Red pine is very susceptible, also on Scots and Austrian. Five-needled pines reported to be resistant.

- **Scorch and dieback**

Description: These diseases commonly result from winter damage as well as from poor root function coupled with high temperatures and moisture shortage. In some cases, bacterial infection of the plant is involved. Scorched leaves are brown around the edges and sometimes between the veins. Leaf yellowing and wilting may occur in late summer. Dieback and decline is often mistaken for Verticillium wilt. No wood streaking is present, however. Girdling roots may also cause these symptoms.

- **Salt burn (chemical)**

Management Options: White pines are quite sensitive to salt spray.

- **Seasonal needle drop (physiological)**

Fig. Road Salt Damage on Pine

Description: In late September and October, the inner and older needles often turn yellow and drop in great numbers. This is a normal process, but often causes homeowner concerns. Occurs on all pines but especially noticeable on white pine. A healthy white pine should retain three years of needles in summer and two years in winter.

Management Options: Plant in a well-drained site, water during extended dry periods, and follow other practices to avoid stress. This will prevent the two-year needles from yellowing and falling a year before they should.

- **Tip blight (fungus)**

Fig. Seasonal Needle Yellowing on Pine

Fig. Diplodia Tip Blight on Pine

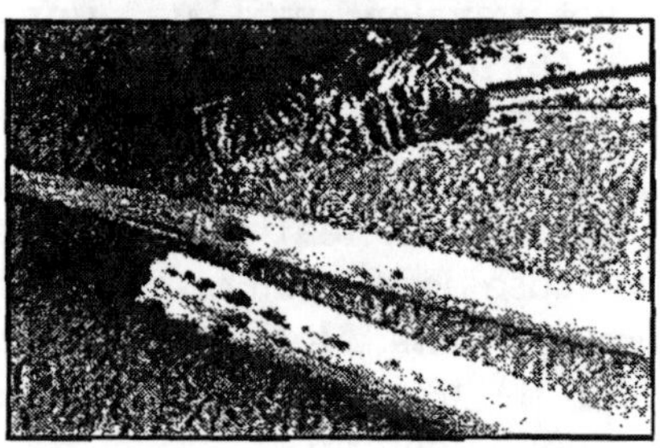

Fig. Close-up of Pine Needles Affected with Diplodia Tip Blight

Description: Dieback of tips of Austrian, red, mugo and Scots pine. Disease often visible early in the season. Stem tips are killed before needles reach full size. Small, black fruiting bodies develop at the base of needles and on the old cones. Resin may ooze from the diseased stem tips. Usually occurs on lower branches first.

Management Options: Where feasible, removal and disposal of diseased twigs may be helpful. If fungicides are used, spray just before new growth starts, followed by one or two more sprays at 14-day intervals. Use a properly labeled fungicide. Fertilize and water trees when needed.

- **White pine root decline (fungus)**

Description: Bud break in the spring is delayed. Needle growth is retarded and may be yellowish. Trees die and needles are reddish brown and do not drop. A canker forms at base of tree. The inner bark is brown to olive-brown.

Management Options: No control. Plant white pines in well-drained sites.

Poplar, Cottonwood, Aspens (Populus)

- **Canker and dieback (fungi)**

Description: Cankers develop on stems, twigs, and trunk. Dieback continues until trees are killed.

Management Options: Lombardy poplar is often killed by canker-causing fungi. Lombardy poplar is very susceptible and should not be planted. Prune and destroy branch cankers.

- **Leaf rust (fungi)**

Description: Orange, powdery pustules on underside of leaves. Usually not prevalent until late summer.

Management Options: Common on some species, others not affected. Rarely affects health of the tree.

- **Leaf spots (fungi)**

Description: Several leaf spot diseases have been identified. Some cause defoliation on susceptible species.

Management Options: Not usually serious enough to threaten the life of the tree. Collecting and destroying fallen diseased leaves may be helpful.

Privet (Ligustrum)

- **Powdery mildew (fungus)**

Description: White, powdery areas appear on leaves,

stems, flowers, or buds. If the infection is widespread, the entire leaf blade may be covered with a thin, white fungal, powdery growth. Some mildew fungi affect older leaves first, such as on oaks, lilacs, zinnias, or chrysanthemums. Others affect newer shoots, such as on roses, flowering dogwood, London plane tree, or crabapples. When new shoots are affected, leaf curling and shoot stunting and twisting may be severe.

Management Options: Control usually not necessary.

- **Twig blight and canker (fungus)**

Description: Stems dieback and pinkish or salmon-colored fungus fruiting bodies develop on the stems.

Management Options: European or common privet is susceptible. Other species reported to have resistance. Prune and destroy diseased stems.

Purple leaf plum and other Prunus (Prunus)

- **Black knot (fungus)**

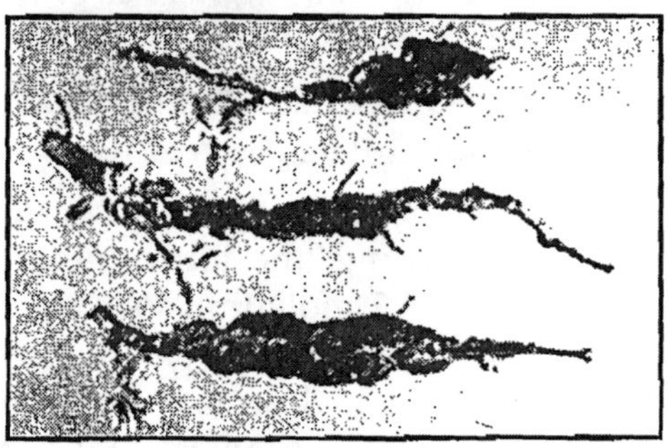

Fig. Black Knot on Plum Twigs

Description: Black, rough galls develop on the twigs. Neglected trees are especially subject to this disease. Black knot is found on some plums and some cherries, not on flowering almond.

Management Options: Prune knotted twigs from large branches during the winter. If necessary, spray with a properly labeled fungicide. Increase the vigour of plants with fertilization if needed. Apply deep root watering during dry spells.

- **Blossom and twig blight and dieback (fungus)**

Description: Blossoms turn brown and collapse. Leaves and twigs become infected and fruits rot. Branches then die back from the infection outward. In damp weather, small growths or masses of powdery, brown fungus spores can be seen over the canker or actual site of the infection, and on the rotting fruit. Flowering almond is very susceptible to dieback.

Management Options: Prune and destroy infected branches. Rake and destroy fallen plant debris. Spray with a properly labeled fungicide just before blossoms open and again 10 days later.

- **Leaf curl (fungus)**

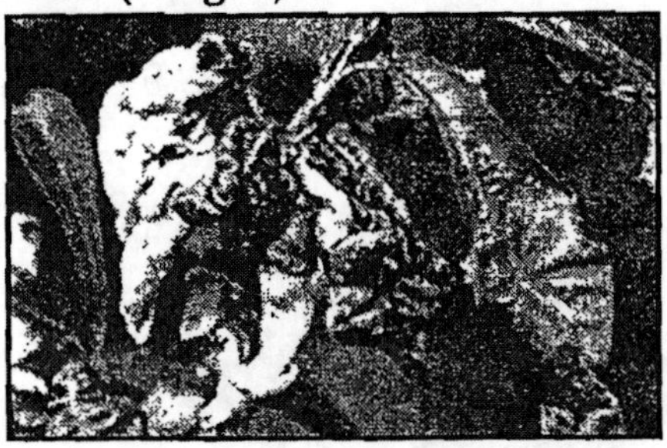

Fig. Leaf Curl on Peach Leaves

Description: Ornamental peaches are commonly affected by this disease. In the spring, the leaves emerge thickened, misshapen, and reddened. Defoliation follows. Fruit- bearing species of peach are very susceptible, as are some native wild cherries. Some species cause malformation on stems and fruits.

Management Options: Spray in late fall, after all leaves have dropped, or in the early spring before buds swell. One properly timed spray should be sufficient. Use a properly labeled fungicide.

- **Leaf spot (fungus)**

Description: Spots often have reddish borders. On cherries and cherry laurel, spots turn brown and centers fall out as they age. Do not confuse this disease with bacterial shot-hole. Worse in wet seasons. Leaf drop may be extensive.

Management Options: White, powdery areas appear on leaves, stems, flowers, or buds. If the infection is widespread, the entire leaf blade may be covered with a thin, white fungal, powdery growth. Some mildew fungi affect older leaves first, such as on oaks, lilacs, zinnias, On some very susceptible Prunus species or cultivars, fungicides may be necessary.

- **Powdery mildew (fungus)**

Description: White, powdery areas appear on leaves, stems, flowers, or buds. If the infection is widespread, the entire leaf blade may be covered with a thin, white fungal, powdery growth. Some mildew fungi affect older leaves first, such as on oaks, lilacs, zinnias.

Management Options: Seldom serious on plums, but can be on some of the cherries.

- **Shot-hole leaf spot (bacteria)**

Description: Small spots are found evenly over the leaf surface. The infected tissue dries up and falls out, leaving holes about 1/8 inch in diameter. The holes are generally smaller than those described under fungal leaf spot (above). Ornamental cherries are particularly susceptible.

Management Options: Plant the trees in sunny locations. Prune away shade-producing vegetation. Do not crowd plants.

Redbud (Cercis)

- **Canker (fungus)**

Description: Sunken, black stem and twig lesions; leaves wilt and turn brown as the stems are girdled. Branch dieback can be extensive. Young twigs and branches die from the terminals and appear burned.

Branches may be bent over at the tip, resembling a shepherd's crook. Dead leaves generally remain on the branch. Affects members of the rose family.

Management Options: Winter injury and drought stress may result in increased susceptibility. Avoid wounding; fertilize and water when necessary.

- **Verticillium wilt (fungus)**

Description: Wilting and yellowing of foliage is followed by premature defoliation. One limb of the tree may be affected first. Gradual dieback may be apparent. Yellow-brown, brown, black, or greenish-black streaks may be found in outer rings of wood of infected branches. In herbaceous plants, the vascular system of stems and roots are discolored.

Rhododendron (Rhododendron)

- Chlorosis, yellowing (Physiological)

Fig. Chlorosis Due to Iron Deficiency on Rhododendron Leaf

Description: Caused by theoil pH being too high (above 6.5)

Management Options: Lower soil pH.

- **Dieback (fungus)**

Description: Caused by the fungus, Botryosphaeria. First symptom usually is drooping leaves on an individual branch. Leaves are rolled inward. A reddish-brown sunken canker girdles the stem.

Management Options: Prune out dying stems, following accepted practices. Fungicides are not effective. This disease will be worse if plants are under stress from winter injury and drought. Keep the plants mulched.

- **Dieback (fungus)**

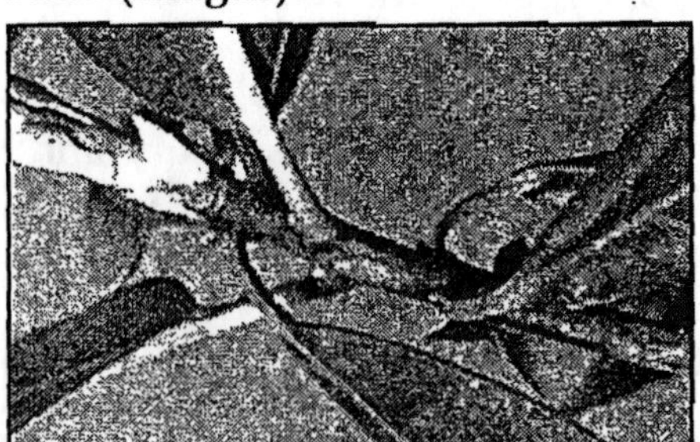

Fig. Phytophthora Stem Canker and Dieback of Rhododendren

Description: Caused by a species of Phytophthora that invades the stems or leaves. Brown leaf lesions are an early symptom. Can be confused with the dieback disease described above.

Management Options: No effective control, other than pruning. Maintain mulch.

- **Powdery mildew (fungus)**

Description: White, powdery areas appear on leaves, stems, flowers, or buds. If the infection is widespread, the entire leaf blade may be covered with a thin, white fungal, powdery growth. Some mildew fungi affect older leaves first, such as on oaks, lilacs, zinnias.

Management Options: More of a problem on some cultivars of azalea. Control not usually necessary on rhododendrons.

- **Root and crown rot (fungus)**

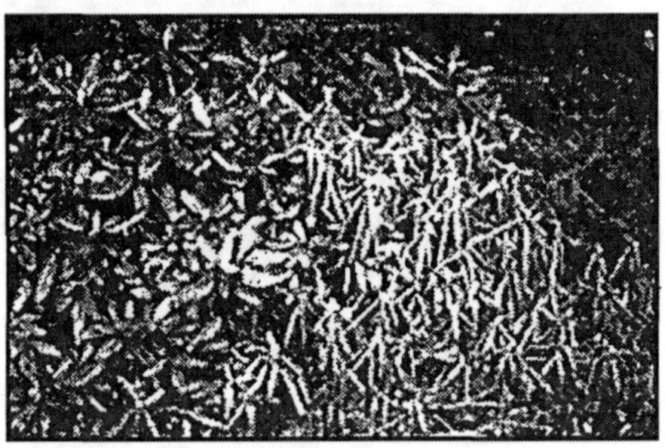

Fig. Above Ground Symptoms of Phytophthora Root and Crown Rot on Rhododendron

Description: Caused by one of several species of water mold fungi. Reddish-brown discoloration of roots and of the stem at or below the soil line. Entire plant wilts.

Management Options: Planting site should be well-drained. Keep the plants mulched properly.

- **Black spot (fungus)**

Description: Black, round to irregular spots with fringed margins mainly on the upper leaf surface. Tissue around the spots may turn yellow. Defoliation often occurs. Small purplish spots occur on the canes.

Management Options: Destroy infected leaves and canes of the previous year. Use an effective spray programme with a properly labeled fungicide beginning when leaves emerge in the spring.

Avoid overhead irrigation, especially late in the day or at night. Prune away shade-producing vegetation to improve rapid drying of foliage. Some varieties have been reported to be resistant or tolerant to black spot.

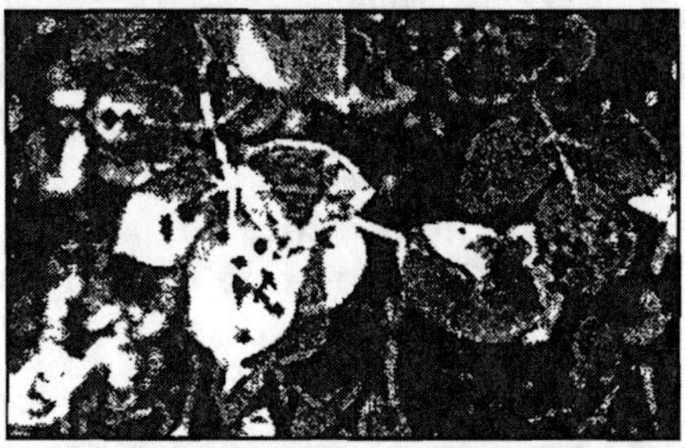

Fig. Black Spot on Rose Leaves

- **Brown canker (fungus)**

Fig. Cane Canker on Rose Cane

Description: Small, reddish-purple spots occur on stems, petioles, leaves, and petals. Stem cankers later turn white and enlarge the following spring into oval, light cinnamon areas studded with small, black, protruding spore-bearing pustules. Stems sometimes are girdled with cankers.

Management Options: Prune out all infected canes and remove or destroy them. Heavy pruning may be necessary if many cankers are present.

- **Botrytis blight (fungus)**

Fig. Botryis Blight on Rose Petal

Description: Buds and partially opened flowers turn brown and decay and are often covered with a powdery gray mold. Irregular tan or brown areas are found on the leaves, especially along the major veins of the leaf blade. The condition usually is noticed in late spring on lower parts of the plant. Deformation and twisting of leaves often result from infections.

Management Options: Seldom serious enough to require control.

- **Crown gall (bacterium)**

Description: Infected plants do not flower well. Galls form at the crown or at the point of grafting.

Management Options: Remove and burn severely infected plants. Keep plants vigorous with fertilizer and water as needed.

- **Powdery mildew (fungus)**

Fig. Powdery Mildew on Rose Leaves

Description: Younger leaves often affected first, often becoming curled.

Management Options: Can occur when rainfall is low or absent. Humidity high at night and low in the day is favorable for development of this disease. Fungicides may be necessary. Some cultivars are reported to be resistant.

Fig. Powdery Mildew on Rose Stem

- **Rust (fungus)**

Description: White, powdery areas appear on leaves, stems, flowers, or buds. If the infection is widespread, the entire leaf blade may be covered with a thin, white fungal, powdery growth. Some mildew fungi affect older leaves first, such as on oaks, lilacs, zinnias.

- **Stem canker or graft canker (fungus)**

Description: Small, pale yellow or reddish spots that increase in size, later turning brown, occur on bark usually near the ground. The cankered wood dries; bark cracks; and small, black fruiting bodies dot the cankered area. Infection enters through grafting wounds, cut stubs of canes, or other wounds.

Management Options: Use measures outlined under "Brown canker." When pruning, cut back to nodes; do not leave stubs.

Russian Olive (Eleagnus)

- **Canker and twig blight (fungus Phomopsis)**

Description: Elongated, reddish-brown stem cankers kill the stems, resulting in dieback. Young twigs and branches die from the terminals and appear burned. Branches may be bent over at the tip, resembling a shepherd's crook. Dead leaves generally remain on the branch. Affects members of the rose family.

Management Options: Avoid wounding the branches. Follow accepted pruning practices. Do not prune when trees are wet or immediately after rains.

- **Verticillium wilt (fungus)**

Description: Wilting and yellowing of foliage is followed by premature defoliation. One limb of the tree may be affected first. Gradual dieback may be apparent. Yellow-brown, brown, black, or greenish-black streaks may be found in outer rings of wood of infected branches. In herbaceous plants, the vascular system of stems and roots are discolored.

Serviceberry (Amelanchier)

- **Fire blight (bacterium)**

Description: Young twigs and branches die from the terminals and appear burned. Branches may be bent over at the tip, resembling a shepherd's crook. Dead leaves generally remain on the branch. Affects members of the rose family.

Management Options: Occasional; usually not serious.

- **Leaf spot (fungus)**

Description: White, powdery areas appear on leaves, stems, flowers, or buds. If the infection is widespread, the entire leaf blade may be covered with a thin, white fungal, powdery growth. Some mildew fungi affect older leaves first, such as on oaks, lilacs, zinnias.

- **Powdery mildew (fungus)**

Description: White, powdery areas appear on leaves,

stems, flowers, or buds. If the infection is widespread, the entire leaf blade may be covered with a thin, white fungal, powdery growth. Some mildew fungi affect older leaves first, such as on oaks, lilacs, zinnias.

Management Options: Control not usually necessary.

- **Rust (fungus)**

Description: White, powdery areas appear on leaves, stems, flowers, or buds. If the infection is widespread, the entire leaf blade may be covered with a thin, white fungal, powdery growth. Some mildew fungi affect older leaves first, such as on oaks, lilacs, zinnias.

Management Options: Control not usually necessary; not common.

Spirea (Spiraea)

- **Fire blight, (bacterium)**

Description: Young twigs and branches die from the terminals and appear burned. Branches may be bent over at the tip, resembling a shepherd's crook. Dead leaves generally remain on the branch. Affects members of the rose family. Reported, but not common.

- **Leaf spot (fungus)**

Description: Small round spots on some species, but not common.

Management Options: Control not necessary unless watered overhead or in wet summers.

Spruce (Picea)

- **Canker (fungus Cytospora)**

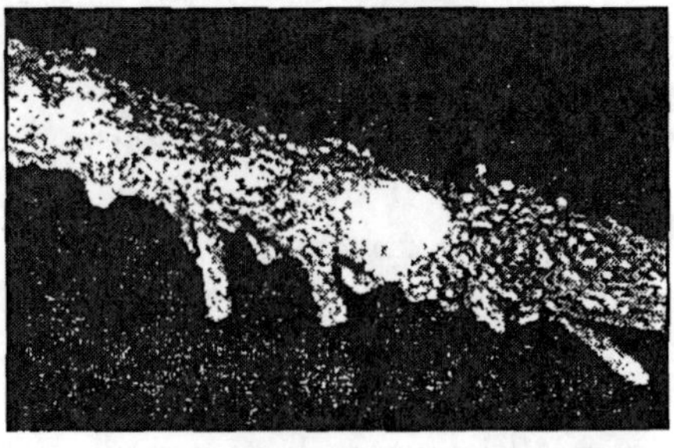

Fig. Close-up of Cytospora on Spruce Twig Canker

Fig. Overall Appearance of a Spruce Tree Affected with Cytospora

Description: First symptoms are browning of needles and dying of selected lower branches. Cankers occur on branches inside crown near the trunk. Swellings, accompanied by white, oozing pitch, occur at the canker. The fungus spreads to higher branches. It is most common on trees more than 15 years old.

Management Options: Prune affected branches at trunk of tree when no rain is expected. Fertilize and water as needed. Vertically mulch if needed to improve the root health. Wounds caused by insects or mechanical injuries appear to be necessary for infection.

Also drought stress is a predisposing factor. Serbian spruce reported to be resistant.

- **Needlecast (fungus Rhizospaera)**

Description: One-year-old needles discolor and have very small, black, fungal fruiting bodies in rows on the needles. Needles fall. Spruce needlecast can also result from soil compaction, heat stress, or herbicide toxicity.

Management Options: Most frequent on blue spruce, not common on Norway spruce.

Sumac

- **Leaf spot (fungus)**

Description: Brown spots with purple-brown borders. Spore-bearing structures visible in the centre of the spot.

Management Options: Usually not serious enough to require control with fungicides.

- **Powdery mildew (fungus)**

Description: White, powdery areas appear on leaves, stems, flowers, or buds. If the infection is widespread, the entire leaf blade may be covered with a thin, white fungal, powdery growth. Some mildew fungi affect older leaves first, such as on oaks, lilacs, zinnias.

Management Options: No control necessary.

- **Verticillium wilt (fungus)**

Description: Wilting and yellowing of foliage is followed by premature defoliation. One limb of the tree may be affected first. Gradual dieback may be apparent. Yellow-brown, brown, black, or greenish-black streaks may be found in outer rings of wood of infected branches. In herbaceous plants, the vascular system of stems and roots are discolored.

Management Options: Not common.

- **Canker (fungus)**

Fig. Botryospharia Canker on Sweet Gum

Description: Cankers develop, girdling the stems, causing dieback. Fruiting bodies of the fungus develop in the cankered bark. "Bleeding" around the cankered bark has been reported in some cases.

Management Options: Fertilize and water trees as necessary to keep them in a vigorous growing condition.

- **Chlorosis or leaf yellowing**

Description: Yellowing of leaves between the veins, the veins remaining green.

Management Options: Often caused by unavailability of iron in high pH soils.

Sycamore (Platanus) and London Plane Tree (Platanus)

- **Anthracnose (fungus)**

Fig. Anthracnose on Sycamore Leaf

Description: New leaves may be killed. Damage looks similar to frost injury. Older or mature leaves show small to large, irregular dead areas along the veins. In the spring, bud blight and twig cankers occur with some dieback.

The lower portions of the tree are damaged more severely. Crooked branching and a witches broom-like proliferation of twigs are also common symptoms. Defoliation occurs after leaf blight in the spring. New leaves develop by summer.

Management Options: Spring sprays are often ineffective because bud infection can occur during the previous autumn. Some cultivars of London plane tree are reported to be less susceptible. Cool temperatures at bud-break are important in affecting severity of this disease.

- Powdery mildew (fungus)

Description: White, powdery areas appear on leaves, stems, flowers, or buds. If the infection is widespread, the entire leaf blade may be covered with a thin, white fungal, powdery growth. Some mildew fungi affect older leaves first, such as on oaks, lilacs, zinnias, On new growth during summer, causing leaf distortion.

Management Options: Control typically not necessary or feasible.

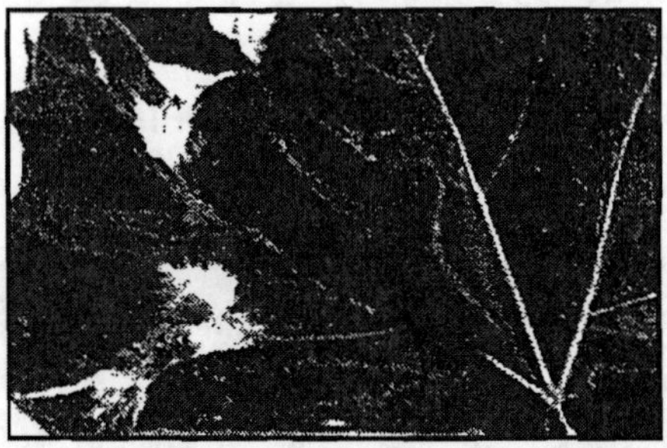

Fig. Powdery Mildew on Sycamore Leaf

Fig. Powdery Mildew on London Plane Tree

- **Scorch and dieback (physiological)**

Description: These diseases commonly result from winter damage as well as from poor root function coupled with high temperatures and moisture shortage. In some cases, bacterial infection of the plant is involved. Scorched leaves are brown around the edges and sometimes between the veins. Leaf yellowing and wilting may occur in late summer. Dieback and decline is often mistaken for Verticillium wilt. No wood streaking is present, however. Girdling roots may also cause these symptoms.

Tulip Poplar (Liriodendron)

- **Leaf spotting and yellowing (physiological)**

Description: Leaves develop large brown to black spots, turn yellow and fall during late-July and August when trees are subject to water stress because of planting site problems or hot, dry weather.

Viburnum (Viburnum)

- **Bacterial leaf spot (bacterium)**

Fig. **Bacterial Leaf Spot on Viburnum Leaves**

Description: Small, circular to angular, brown, water-soaked spots. Spots become irregular.

Management Options: Not common or rare on most landscape species. Has been noted on European cranberrybush, Burkwood, viburnum, Koreanspice viburnum, and Juddii viburnum. Worse during wet season or when using overhead watering.

- **Downy mildew (fungus)**

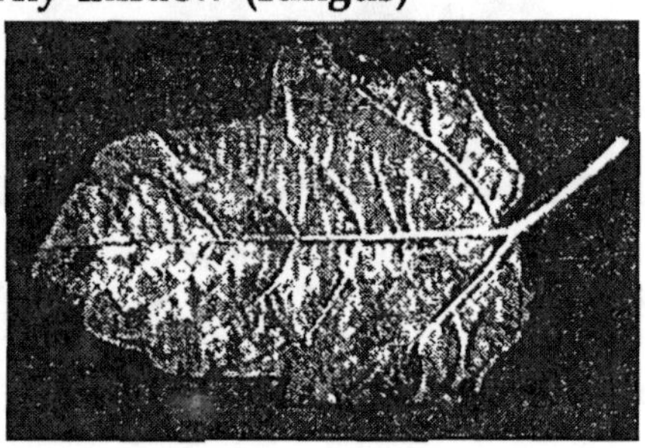

Fig. Downy Mildew on Viburnun Leaf

Description: Certain species will develop gray, sometimes water-soaked spots on the underside of leaves, with brownish, scorched areas in the upper leaf surface. As the disease worsens, defoliation can be severe. Common in wet seasons or if watered overhead.

Management Options: Prune away shade-producing vegetation. Do not crowd the plants. Avoid overhead irrigation. Spray with a properly labeled fungicide. *V. opulus* is particularly susceptible. Not common on most other species.

- **Powdery mildew (fungus)**

Description: White, powdery areas appear on leaves, stems, flowers, or buds. If the infection is widespread, the entire leaf blade may be covered with a thin, white fungal, powdery growth. Some mildew fungi affect older leaves first, such as on oaks, lilacs, zinnias.

Management Options: A common problem on *V. lentago* and *V. prunifolium*. Rare or not found on other landscape species.

- **Anthracnose (fungus)**

Fig. Anthracnose on Walnut Leaf

Description: Irregular dark brown spots on leaflets. Yellowing occurs around the spots, and leaf drop occurs in late summer. Leaf fall is often extensive.

Management Options: Since this is a disease that occurs late in the summer, little actual damage to the tree occurs. May be feasible to spray small trees.

Yew (Taxus)

- **Chlorosis or yellowing (several potential causes)**

Description: Young twigs and branches die from the

terminals and appear burned. Branches may be bent over at the tip, resembling a shepherd's crook. Dead leaves generally remain on the branch. Affects members of the rose family.

Management Options: Do not plant in poorly drained soils. Examine for black vine weevil feeding on leaves, roots, and lower stem. If black vine weevil is present, use chemical controls.

- **Root rot (fungus)**

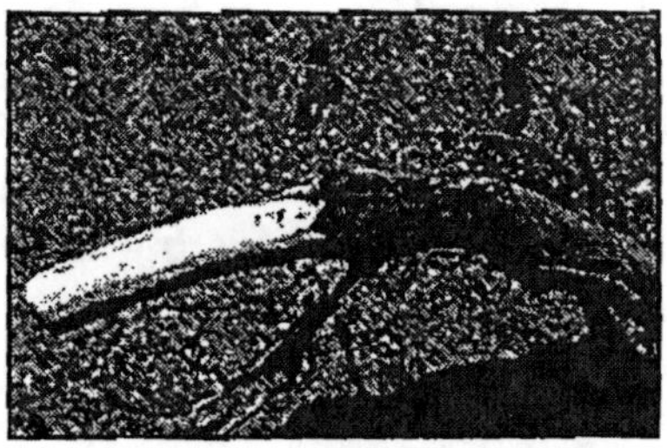

Fig. Taxus Root Affected with Phytophthora Root Rot

Description: These diseases commonly result from winter damage as well as from poor root function coupled with high temperatures and moisture shortage.

In some cases, bacterial infection of the plant is involved. Scorched leaves are brown around the edges and sometimes between the veins.

Leaf yellowing and wilting may occur in late summer. Dieback and decline is often mistaken for Verticillium wilt. No wood streaking is present, however. Girdling roots may also cause these symptoms.

Management Options: Plant in well-drained sites or improve drainage if possible.

Zelkova (Zelkova)

- **Dutch elm disease (fungus)**

Management Options: Zelkova is only mildly susceptible to Dutch elm disease. Prompt pruning of infected branches is often an effective control.

- **Canker (fungus)**

Description: Young twigs and branches die from the

terminals and appear burned. Branches may be bent over at the tip, resembling a shepherd's crook. Dead leaves generally remain on the branch. Affects members of the rose family.

SPECIFIC DISEASES OF HERBACEOUS ANNUALS AND PERENNIALS

Aster (Aster) Hardy Aster

- **Botrytis stem rot**

Description: Irregular tan or brown areas are found on the leaves, especially along the major veins of the leaf blade. The condition usually is noticed in late spring on lower parts of the plant. Deformation and twisting of leaves often result from infections.

- **Fusarium wilt (fungus)**

Description: Look for dark streaks up one side of the stem on older plants. Death of the entire plant may soon follow initial symptoms.

Base of the stem is dark with a brown vascular ring just beneath the surface. Roots will be rotted. Pink fungus spores may be seen at the base of the stem.

Management Options: Grow plants in a new location each year. Maintain plant vigour with appropriate watering, fertilization, etc. Not serious on the hardy asters. China asters susceptible.

- **Powdery mildew (fungus)**

Description: White, powdery areas appear on leaves, stems, flowers, or buds.

If the infection is widespread, the entire leaf blade may be covered with a thin, white fungal, powdery growth. Some mildew fungi affect older leaves first, such as on oaks, lilacs, zinnias.

Management Options: Control seldom necessary.

- **Rust (fungus)**

Description: White, powdery areas appear on leaves, stems, flowers, or buds. If the infection is widespread, the entire leaf blade may be covered with a thin, white fungal, powdery growth. Some mildew fungi affect older leaves first, such as on oaks, lilacs, zinnias, Yellow-orange pustules on underside of leaf.

Management Options: Destroy goldenrod and wild asters that are weed hosts.

- **Yellows (mycoplasma)**

Description: New growth is yellowish; plants are dwarfed. Upper leaves are erect. Flower heads open unevenly or only partly and have a yellowish-green cast.

Management Options: Pull up and burn diseased plants when noticed. Control weeds, many of which are also hosts. Use insecticides to control leafhoppers that spread the mycoplasma to neighboring asters.

Chrysanthemum (Chrysanthemum)

- **Botrytis blight (fungus)**

Fig. Botrytis Blight on Chrysanthemum Blossom

Description: Irregular tan or brown areas are found on the leaves, especially along the major veins of the leaf blade. The condition usually is noticed in late spring on lower parts of the plant. Deformation and twisting of leaves often result from infections. Only occasional on garden mums.

- **Foliar nematode (nematode)**

Description: Wedge-shaped portions of leaf turn yellowish-brown between the veins. These run together, causing the leaf to die and fall.

Management Options: Remove infected plants. Do not plant mums for three years in the affected area. Treat soil with a soil fumigant labeled for the particular site.

- **Leaf spot (fungus)**

Description: Small circular spots, at first on the lower leaves, becoming irregular dark brown to black. Leaves shrivel but commonly hang down on the stem.

Management Options: Keep the foliage dry when possible. If the disease is serious, fungicides may have to become part of the control programme. Clean up and destroy plant debris.

Fig. Septoria Leaf Spot on Chrysanthemum

- **Powdery mildew (fungus)**

Description: White, powdery areas appear on leaves, stems, flowers, or buds. If the infection is widespread, the entire leaf blade may be covered with a thin, white fungal, powdery growth. Some mildew fungi affect older leaves first, such as on oaks, lilacs, zinnias, Generally not common or serious on garden mums.

- **Rust (fungus)**

Description: White, powdery areas appear on leaves, stems, flowers, or buds. If the infection is widespread, the entire leaf blade may be covered with a thin, white fungal, powdery growth.

Some mildew fungi affect older leaves first, such as on oaks, lilacs, zinnias, Pale yellow flecks develop on upper leaf surface, followed by brown, powdery pustules on lower leaf surface.

Management Options: Usually of minor importance on chrysanthemums in Midwest gardens.

- **Stunt and virus diseases (viruses and virus-like agents)**

Description: Chlorosis, mosaic, and stunting are some of the symptoms. Flowering earlier than normal.

Management Options: More of a potential problem on greenhouse mums. A virus-indexing programme by the specialist propagator is the key to managing these diseases.

- **Verticillium wilt (fungus)**

Description: Wilting and yellowing of foliage is followed by premature defoliation. One limb of the tree may be affected first. Gradual dieback may be apparent. Yellow-brown, brown, black, or greenish-black streaks may be found in outer rings of wood of infected branches. In herbaceous plants, the vascular system of stems and roots are discolored.

Management Options: Some chrysanthemum cultivars are resistant, but no up-to-date list is available. Not a common problem in recent years.

Columbine (Aquilegia)

- **Crown and root rot (fungus)**

Description: Plants wilt, white mats or cottony patches of fungus often visible at base of plant; small hard, dark bodies are also visible at base of plant and within lower stem when cut lengthwise. These diseases commonly result from winter damage as well as from poor root function coupled with high temperatures and moisture shortage.

In some cases, bacterial infection of the plant is involved. Scorched leaves are brown around the edges and sometimes between the veins. Leaf yellowing and wilting may occur in late summer.

Dieback and decline is often mistaken for Verticillium wilt. No wood streaking is present, however. Girdling roots may also cause these symptoms.

Management Options: Soil drenches may be helpful. These fungi survive in the soil for long periods of time. Plant in a new location.

Coreopsis (Coreopsis) (Tickseed)

Powdery mildew (fungus)

Description: White, powdery areas appear on leaves,

stems, flowers, or buds. If the infection is widespread, the entire leaf blade may be covered with a thin, white fungal, powdery growth. Some mildew fungi affect older leaves first, such as on oaks, lilacs, zinnias.

Management Options: Control seldom necessary.

- **Stem rot (fungus)**

Description: Plants wilt, lower stem discolored. White mats or cottony patches of fungus visible at base of plant. Hard, black bodies visible on or within the stem.

Management Options: Not very common. The fungus may persist for long periods of time in the soil. If serious, plant in a new location.

Dahlia (Dahlia)

- **Botrytis blight (fungus)**

Description: Irregular tan or brown areas are found on the leaves, especially along the major veins of the leaf blade. The condition usually is noticed in late spring on lower parts of the plant. Deformation and twisting of leaves often result from infections that occ Flowers, leaves, and stems are affected especially under conditions of high humidity for prolonged periods.

Management Options: Remove and destroy infected or old flowers.

- **Botrytis blight (fungus)**

Description: Irregular tan or brown areas are found on the leaves, especially along the major veins of the leaf blade. The condition usually is noticed in late spring on lower parts of the plant. Deformation and twisting of leaves often result from infections that occ Flowers, leaves, and stems are affected especially under conditions of high humidity for prolonged periods.

Management Options: Remove and destroy infected or old flowers.

- **Mosaic stunt or dwarfing (viruses)**

Description: A common characteristic is stunt or dwarfing with a tendency to produce many shortened lateral shoots. A bushy plant with mottled leaf colour is produced. Scattered areas of yellow or light green tissue occur in the leaves at first. Conspicuous ring spots develop on leaves in some cases.

Management Options: Dig up and remove or destroy diseased plants. Plant only tubers from healthy plants. Control aphids or other vectors with approved insecticides.

- **Powdery mildew (fungus)**

Description: White, powdery areas appear on leaves, stems, flowers, or buds. If the infection is widespread, the entire leaf blade may be covered with a thin, white fungal, powdery growth. Some mildew fungi affect older leaves first, such as on oaks, lilacs, zinnias.

Management Options: No treatment usually necessary for powdery mildew.

- **Ring spot (virus)**

Description: Scattered areas of yellow or light green tissue occur in the leaves at first. A conspicuous ring pattern develops later on the leaves.

Management Options: Control thrips with approved insecticides. Thrips spread the virus to neighboring plants. Plant only tubers from healthy plants.

Delphinium (Delphinium)

- **Black leaf spot (bacterium)**

Description: Black, irregular spots occur on upper surfaces of leaves, petioles, stems, and flowers; the lower leaf surfaces have brown spots. Lower leaves are infected first. Disease gradually progresses upward until the entire stalk is killed.

Management Options: Cut off and burn old tops. Avoid overhead irrigation. Prune away shade-producing vegetation. If necessary, spray with properly labeled material.

- **Bud and crown rot (bacterium)**

Description: Infection results in a rapid wilting of the whole plant and finally death. The decay is characterized by a strong, offensive odor.

Management Options: Avoid planting in low, poorly drained areas. Avoid overhead irrigation, especially late in the day. Control soil insects in affected areas.

- **Powdery mildew (fungus)**

Description: White, powdery areas appear on leaves, stems, flowers, or buds. If the infection is widespread, the entire leaf blade may be covered with a thin, white fungal, powdery growth. Some mildew fungi affect older leaves first, such as on oaks, lilacs, zinnias.

Management Options: Control not usually necessary.

- **Root, stem or crown rot (fungi)**

Description: Two different fungi can cause this disease and kill the plant. In both cases white strands or mats of fungus become visible on and around the base of the plants. Large, hard, black fungus bodies soon form on or in the stems. A second fungus forms small, round, seed-like, tan to dark brown bodies on the white fungal strands and stems of diseased plants.

False dragonhead (Physostegia)

- **Crown and root rots (fungi)**

Description: Plants wilt and die. White, fluffy strands of fungus are visible at base of the plant. One fungus produces hard, black bodies on the outer stems or inside. They can be observed when stems are cut lengthwise. Another fungus produces small, round, seed-like bodies. These are cream-colored, becoming tan and brown.

Management Options: These diseases commonly result from winter damage as well as from poor root function coupled with high temperatures and moisture shortage. In some cases, bacterial infection of the plant is involved. Scorched leaves are brown around the edges and sometimes between the veins. Leaf yellowing and wilting may occur in late summer. Dieback and decline is often mistaken for Verticillium wilt. No wood streaking is present, however. Girdling roots may also cause these symptoms.

Fuchsia (Fuchsia)

- **Botrytis (fungus)**

Description: Irregular tan or brown areas are found on the leaves, especially along the major veins of the leaf blade.

The condition usually is noticed in late spring on lower parts of the plant. Deformation and twisting of leaves often result from infections that occ Causes a leaf and stem rot.

Management Options: Develops under high humidity. Remove badly blighted parts.

- **Rust (fungus)**

Description: White, powdery areas appear on leaves, stems, flowers, or buds. If the infection is widespread, the entire leaf blade may be covered with a thin, white fungal, powdery growth. Some mildew fungi affect older leaves first, such as on oaks, lilacs, zinnias.

Management Options: Not common. When observed, it usually is on plant material brought to the garden centre from another location.

Garden Lily (Lilium)

- **Botrytis blight (fungus)**

Description: Symptoms are numerous circular to oval leaf spots. They are light brown and are surrounded by a reddish-purple margin. Lesions also occur on buds, flowers, and stems. Bulbs may be affected.

Management Options: Destroy diseased plant parts. Plant good- quality bulbs.

- **Stem rot (fungi)**

Description: Two different water mold fungi cause stem rot. One causes rot of the lower stem, and another rots the stem near the top of the plant and is called top rot.

Management Options: These diseases commonly result from winter damage as well as from poor root function coupled with high temperatures and moisture shortage. In some cases, bacterial infection of the plant is involved. Scorched leaves are brown around the edges and sometimes between the veins. Leaf yellowing and wilting may occur in late summer. Dieback and decline is often mistaken for Verticillium wilt. No wood streaking is present, however. Girdling roots may also cause these symptoms.

- **Virus diseases (viruses)**

Description: Symptoms are flecking, necrotic streaks, and mottling of the leaves; plants may be stunted, flowers distorted or subnormal Several different viruses can affect a single plant.

Management Options: Remove and destroy diseased plants. Keep insect vectors (aphids) under control. More of a problem on Easter lily in the greenhouse.

Geranium (Pelargonium)

- **Bacterial blight (bacterium)**

Description: Small, water-soaked spots, later becoming brown. Yellow patches and wedge-shaped brown areas develop. Leaves wilt or yellow. Later, stems develop internal vascular discoloration and finally a brown to black stem rot.

Management Options: Purchase symptom-free plants. Sanitation and other practices by the greenhouse propagators are important in providing disease-free plants for the garden.

- **Botrytis gray mold (fungus)**

Fig. Botrytis Blight on Geranium Flower

Fig. Botrytis Blight on Geranium Leaf

Description: Leaf and blossom blight are the symptoms. Irregular tan or brown areas are found on the leaves, especially along the major veins of the leaf blade. The condition usually is noticed in late spring on lower parts of the plant. Deformation and twisting of leaves often result from infections.

- **Rust (fungus)**

Fig. Symptoms of Geranium Rust on Upper Leaf Surface

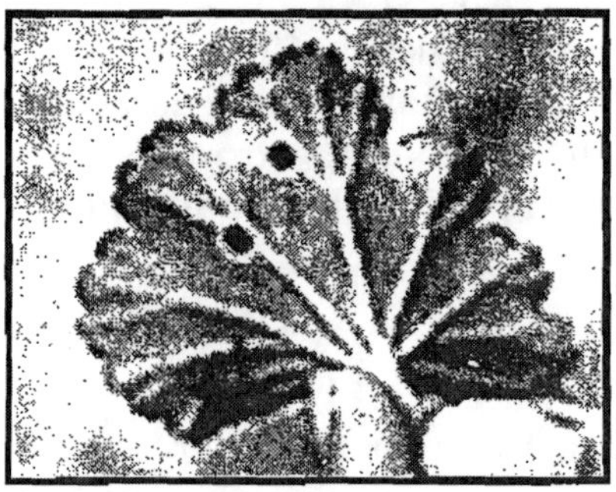

Fig. Symptoms of Geranium Rust on Lower Leaf Surface

Description: Rust-colored spore pustules in concentric circles form on the lower leaf surface. Yellow spots form on the upper surface.

Management Options: Not common or persistent on plants in the landscape.

- **Stem and root rots (fungi)**

Description: Caused by soil-borne fungi. Roots are invaded and rotted. In some cases, a black rot develops at the base of the stem. Often a problem during propagation when the cuttings are being rooted.

Management Options: Sanitation and use of pathogen-free rooting materials are important.

Gladiolus (Gladiolus)

- **Corm rots (fungi, bacteria)**

Description: Several speciea of fungi and bacteria result in corm lesions and leaf spots.

Management Options: Commercially produced corms should be free of decay. If saving corms, avoid injury when digging. Store in cool, dry locations. Discard badly diseased corms.

- **Fusarium yellows and corm rot (fungus)**

Description: Symptoms are variable. Leaves may yellow. Base of leaves become diseased and the corms rot. Flowering may be delayed or lacking.

Management Options: Some cultivars are tolerant. Inspect corms and discard those with any signs of decay or rot.

- **Mosaic (viruses)**

Description: Mottling of leaves, whitish streaks, and colour abnormality in flowers are characteristics of mosaic.

Management Options: Destroy diseased plants. Control insects, especially aphids, which transmit the virus to healthy plants.

Hollyhock (Alcea)

- **Anthracnose blight (fungus)**

Description: Irregular tan or brown areas are found on the leaves, especially along the major veins of the leaf blade. The condition usually is noticed in late spring on lower parts of the plant. Deformation and twisting of leaves often result from infections.

Management Options: Anthracnose leaf blight and other fungal leaf spots are reported but are seldom serious.

- **Root, stem or crown rot (fungus)**

Description: These diseases commonly result from winter damage as well as from poor root function coupled with high temperatures and moisture shortage. In some cases, bacterial infection of the plant is involved. Scorched leaves are brown around the edges and sometimes between the veins. Leaf yellowing and wilting may occur in late summer. Dieback and decline is often mistaken for Verticillium wilt. No wood streaking is present, however. Girdling roots may also cause these symptoms.

- **Rust (fungus)**

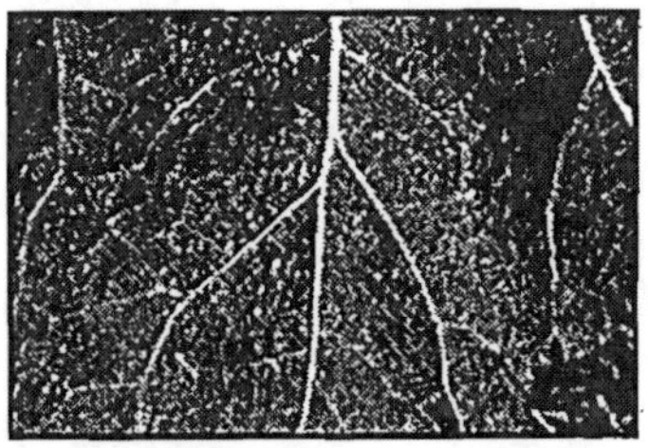

Fig. Early Stages of Hollyhock Rust on Lower Leaf Surface

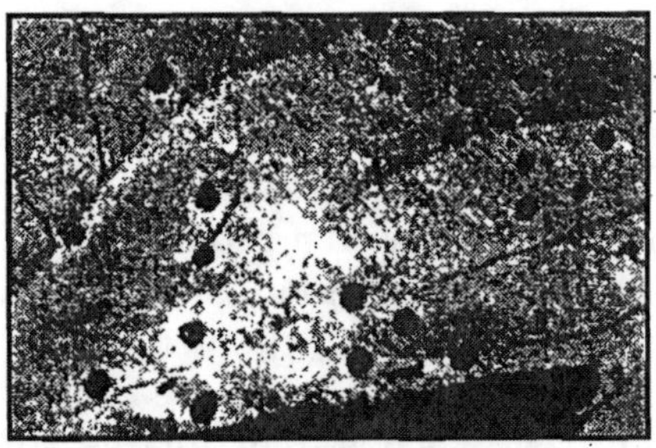

Fig. Later Stages of Hollyhock Rust on Lower Leaf Surface.

Fig. Plant Severely Affected with Hollyhock Rust.

Description: Abundant orange to dark brown pustules on leaf blades, petioles, and stems. White, powdery areas appear on leaves, stems, flowers, or buds.

If the infection is widespread, the entire leaf blade may be covered with a thin, white fungal, powdery growth. Some mildew fungi affect older leaves first, such as on oaks, lilacs, zinnias.

Management Options: This rust is systemic. It survives inside the crown of the plant. Destroy the plant. Purchase rust-free plants.

Iris

- **Crown rot (fungus)**

Description: Tips of outer leaves die. The dead tissue develops until the entire leaf is dead. As rot progresses inward at the base of the plant, the leaves collapse. White fan-like fungus growth may be seen between leaves near the ground line. Rhizomes are not destroyed but are weakened. Light tan to brown fungal structures, the size of mustard seeds, are present on rhizomes and on the base of the plant.

Management Options: Do not permit crowding of plants. Dig out rhizomes and replant only healthy material. Before replanting, treat the area with a properly labeled fungicide or plant in an area not recently planted with iris.

- **Leaf spot (fungus)**

Description: Small, oval spots are brown and often have a water-soaked margin. They may enlarge and blight most of the leaf. Older spots have gray centers and are dotted with small, black spore clusters. White, powdery areas appear on leaves, stems, flowers, or buds. If the infection is widespread, the entire leaf blade may be covered with a thin, white fungal, powdery growth. Some mildew fungi affect older leaves first, such as on oaks, lilacs, zinnias.

Management Options: In the fall, clean up and destroy the leaves. Fungicides may be necessary in some cases.

- **Rust (fungus)**

Description: White, powdery areas appear on leaves, stems, flowers, or buds. If the infection is widespread, the entire leaf blade may be covered with a thin, white fungal, powdery growth. Some mildew fungi affect older leaves first, such as on oaks, lilacs, zinnias.

Management Options: Can be a problem on some species, but not reported on the common rhizomatous garden species.

- **Soft rot (insect followed by bacterium)**

Description: At first leaves wilt slightly, then later become limp and die. The rhizome develops a soft, slimy rot and later turns dry and granular, finally decaying entirely. The problem is always associated with wounds or injury, usually those caused by the iris borer.

Management Options: Do not permit crowding of plants. Dig out infected rhizomes and cut out rotted areas. Treat with insecticide for borer control.

Larkspur, Annual (Delphinium)

- **Root, stem and crown rots (fungi)**

Description: The diseases listed under Delphinium may occur on Larkspur, but most are not common on the Annual Larkspur.

Marigold (Tagetes)

- **Basal stem rot (fungus)**

Fig. Basal Stem Rot of Marigold Caused by Botrytis Cinerea

Description: Plant wilts, sunken and discolored area on lower stem. Roots often rot as well.

Management Options: Caused by a water mold fungus, or sometimes by Botrytis. Not common. Do not crowd plants and plant in a well-drained site.

Monarda

- **Powdery mildew (fungus)**

Description: White, powdery areas appear on leaves, stems, flowers, or buds. If the infection is widespread, the entire leaf blade may be covered with a thin, white fungal, powdery growth. Some mildew fungi affect older leaves first, such as on oaks, lilacs, zinnias.

Management Options: Powdery mildew may be a problem when clumps become too large or crowded. Provide good air circulation.

- **Rust (fungus)**

Description: White, powdery areas appear on leaves, stems, flowers, or buds. If the infection is widespread, the entire leaf blade may be covered with a thin, white fungal, powdery growth. Some mildew fungi affect older leaves first, such as on oaks, lilacs, zinnias.

Management Options: Usually does not require control.

Narcissus (Daffodil)

- **Basal rot (fungus)**

Description: Primarily affects the bulbs, rotting the interior tissues at first, Usually a disease of bulbs in storage. Lightly infected bulbs, if planted, result in stunted plants, yellow leaves, and premature death.

Management Options: If storing bulbs, dry them quickly; avoid wounding when digging and cleaning prior to storage. Discard all infected bulbs. Bulbs from commercial sources should be free of disease.

- **Stem nematodes (nematodes)**

Description: Flower buds fail to open; mis-shaped stems and leaves; plants decline. Bulbs are also affected.

Management Options: Destroy diseased plants and affected bulbs.

Pansy (Viola)

- **Anthracnose and leaf spots (fungus)**

Description: Irregular tan or brown areas are found on the leaves, especially along the major veins of the leaf blade. The condition usually is noticed in late spring on lower parts of the plant. Deformation and twisting of leaves often result from infections that occ Size and colour of spots vary depending on fungus involved.

Management Options: Fungicide sprays may sometimes be necessary.

- **Botrytis blight (fungus)**

Description: Irregular tan or brown areas are found on the leaves, especially along the major veins of the leaf blade. The condition usually is noticed in late spring on lower parts of the plant. Deformation and twisting of leaves often result from infections.

- **Root rot (fungus)**

Description: Older leaves turn yellow. Plants are stunted and may die. Roots rot and may become black. Management **Options:** Destroy diseased plants.

Peony (Paeonia)

- **Bud blast (environmental)**

Description: Bud blast is a condition where flower buds abort. It is caused by planting too deep.

- **Botrytis blight (fungus)**

Fig. Botrytis Leaf Blight on Peony

Description: Irregular tan or brown areas are found on the leaves, especially along the major veins of the leaf blade. The condition usually is noticed in late spring on lower parts of the plant. Deformation and twisting of leaves often result from infections.

Management Options: All parts of the plant may be affected. Sanitary measures are helpful; remove and destroy diseased plant parts.

Fig. Botrytis Stem Blight on Peony

- **Leaf blotch (fungus)**

Description: Glossy, dark purple spots appear on top of the leaf; spots are dull chestnut on the lower surface of the leaf. Worse during moist weather on plants in shaded sites. Leaves are usually not killed.

Management Options: Destroy old infected foliage at the end of season. Avoid overhead irrigation. Prune away surrounding shade-producing vegetation. Spray at first sign of disease with properly labeled fungicide.

- **Root knot (nematode)**

Description: Sprouts are weak and spindly. Larger roots are short, stubby, knotted, and gnarled. Smaller roots contain many small swellings.

Management Options: Discard badly infected plants and plant clean plants in non-infested soil. Treat the soil with a properly labeled nematicide or have soil fumigated. Destroy old plants, and plant new material in a new location.

- **Root, stem or crown rot (fungus)**

Description: These diseases commonly result from winter damage as well as from poor root function coupled with high temperatures and moisture shortage.

In some cases, bacterial infection of the plant is involved. Scorched leaves are brown around the edges and sometimes between the veins. Leaf yellowing and wilting may occur in late summer. Dieback and decline is often mistaken for Verticillium wilt. No wood streaking is present, however. Girdling roots may also cause these symptoms.

Management Options: Remove and destroy all plant

debris. In severe cases the entire plant should be dug and destroyed.

- **Verticillium wilt (fungus)**

Description: Wilting and yellowing of foliage is followed by premature defoliation. One limb of the tree may be affected first. Gradual dieback may be apparent. Yellow-brown, brown, black, or greenish-black streaks may be found in outer rings of wood of infected branches. In herbaceous plants, the vascular system of stems and roots are discolored.

Petunia (Petunia)

- **Botrytis blight (fungus)**

Description: Irregular tan or brown areas are found on the leaves, especially along the major veins of the leaf blade. The condition usually is noticed in late spring on lower parts of the plant. Deformation and twisting of leaves often result from infections.

- **Root, stem or crown rot (fungi)**

Description: These diseases commonly result from winter damage as well as from poor root function coupled with high temperatures and moisture shortage. In some cases, bacterial infection of the plant is involved. Scorched leaves are brown around the edges and sometimes between the veins. Leaf yellowing and wilting may occur in late summer. Dieback and decline is often mistaken for Verticillium wilt. No wood streaking is present, however. Girdling roots may also cause these symptoms.

Management Options: At least three different fungi may be involved. Plants wilt and die. More likely on plants set too deep. Be sure planting site is well drained.

Phlox (Phlox)

- **Leaf blight (physiological)**

Description: Symptoms first observed on lower leaves that die starting at the base. Found on older plants, especially old stems from which new shoots arise in the spring.

Management Options: The problem is associated with moisture translocation in the plant. Old stems are unable to supply water to the leaves in dry weather on perennial phlox. Mulching may help.

- **Leaf spot (fungus)**

Description: Spots are tan to dark brown and circular with light gray centers. White, powdery areas appear on leaves, stems, flowers, or buds. If the infection is widespread, the entire leaf blade may be covered with a thin, white fungal, powdery growth. Some mildew fungi affect older leaves first, such as on oaks, lilacs, zinnias.

Management Options: Not common and rarely needs control.

- **Powdery mildew (fungus)**

Description: White, powdery areas appear on leaves, stems, flowers, or buds. If the infection is widespread, the entire leaf blade may be covered with a thin, white fungal, powdery growth. Some mildew fungi affect older leaves first, such as on oaks, lilacs, zinnias.

Management Options: Perennial or garden phlox and annual phlox are susceptible. Resistant cultivars may be available. Fungicides may be necessary.

- **Stem nematodes (nematodes)**

Description: Plants decline; stem tips and leaves yellow, then die, and flowering is poor. Winter dieback may be severe.

Management Options: Remove and destroy infected plants. Do not plant phlox back into severely affected beds.

Snapdragon (Antirrhinum)

- **Botrytis blight (fungus)**

Description: Flowers are invaded and eventually the smaller branches are infected, killing the entire flower spike. Irregular tan or brown areas are found on the leaves, especially along the major veins of the leaf blade. The condition usually is noticed in late spring on lower parts of the plant. Deformation and twisting of leaves often result from infections.

Management Options: Remove infected plant parts and thin plants to maintain good air circulation (faster drying of plant parts).

- **Leaf spot and blight (fungi)**

Description: Grayish-white to brown leaf spots depending on the fungus involved. In some cases stem lesions may develop, causing a more extensive blight.

Management Options: Remove badly diseased plants and general sanitation measures should be practiced. Fungicides may be helpful in some situations.

- **Powdery mildew (fungus)**

Description: White, powdery areas appear on leaves, stems, flowers, or buds. If the infection is widespread, the entire leaf blade may be covered with a thin, white fungal, powdery growth. Some mildew fungi affect older leaves first, such as on oaks, lilacs, zinnias.

Management Options: Not common on garden snapdragons.

- **Rust (fungus)**

Description: White, powdery areas appear on leaves, stems, flowers, or buds. If the infection is widespread, the entire leaf blade may be covered with a thin, white fungal, powdery growth. Some mildew fungi affect older leaves first, such as on oaks, lilacs, zinnias.

Management Options: Rust-resistant varieties are available.

- **Stem, crown and root rot (fungi)**

Description: Stems near the ground become gray or discolored; entire plant then wilts. A visible white cottony fungus growth develops on the outside of the stem near the ground. Large, hard, black bodies form on the outside and within the stems. Botrytis also can cause a grayish stem rot near the base of the stem. Also a microscopic water mold fungus causes stem and root rot of snapdragons.

Tulip (Tulipa)

- **Botrytis blight (fungus)**

Description: Young shoots, leaves, stems, flowers, and bulbs may all be affected. Grayish spore masses develop on diseased areas.

Management Options: High humidities and frequent rain periods are favorable conditions for disease development.

Badly affected plants should be destroyed. Do not plant bulbs with abnormal spots.

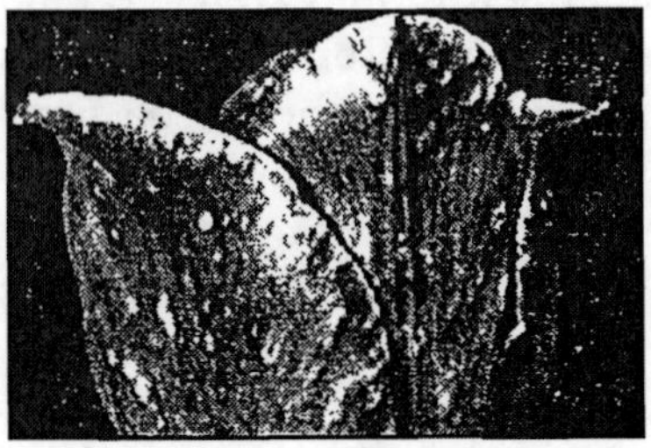

Fig. Symptoms of Botrytis Infections on Tulip Petals

Fig. Symptoms of Botrytis Infections on Tulip Leaves.

- **Basal stem and bulb rots (fungus)**

Description: Plants may not emerge; bulbs rot. Leaves that do develop soon die and rot at soil level.

Management Options: Select clean bulbs for planting. If disease occurs, remove and destroy plants. If the disease has been common, avoid planting in that area.

Zinnia (Zinnia)

- **Alternaria leaf blight (fungus)**

Description: Circular to irregular leaf spots develop that are purple to dark brown and up to 1Ž2 inch in diameter. Centers often become grayish-white. Leaves become brown and dry. Much smaller spots develop on the ray flowers. Girdling stem lesions may occur.

Management Options: Avoid overhead watering. Destroy diseased plants before overwintering.

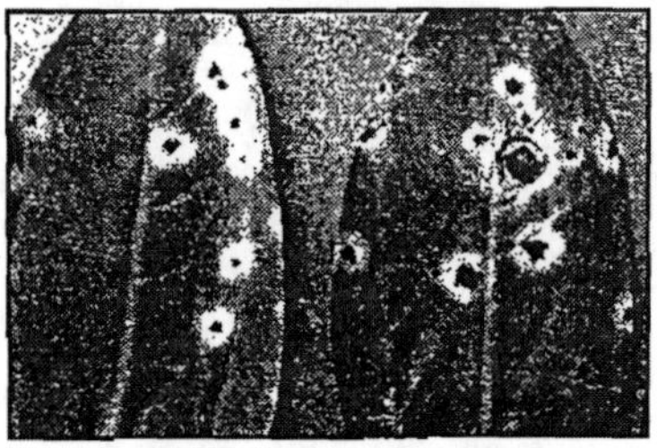

Fig. Alternaria Leaf Blight on Zinnia

- **Bacterial leaf spot (bacterium)**

Fig. Bacterial Leaf Spot on Zinnia

Description: Reddish-brown to dark brown, angular spots up to 1/8 inch in diameter, with a prominent yellow halo. Spots may coalesce.

Management Options: Avoid overhead irrigation. The disease is seed-borne; do not plant diseased seedlings. Destroy diseased plants, especially if infected when young.

- **Powdery mildew (fungus)**

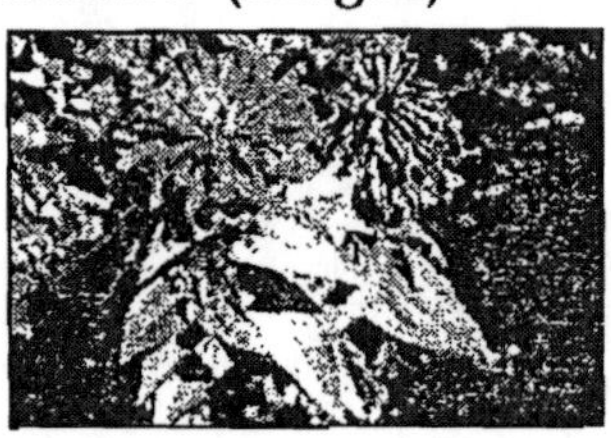

Fig. Powdery Mildew on Zinnia Leaves

Description: Grayish-white, powdery growth on the leaves, in patches at first. Usually develops in late summer and is more severe in shaded locations.

Management Options: Plant in full sun. On problem sites, fungicides may be useful.